ENVIRONMENTAL BIOLOGY

ENVIRONMENTAL BIOLOGY

Edited by
Prof. P. R. TRIVEDI
Chairman
Indian Institute of Ecology and Environment

Prof. GURDEEP RAJ
Hony. Professor in Environmental Science
I.E.E.E., New Delhi
Reader in Chemistry
D. S. College, University of Delhi

AKASHDEEP PUBLISHING HOUSE
NEW DELHI-110002 (INDIA)

AKASHDEEP PUBLISHING HOUSE
4374/4B, Ansari Road, Daryaganj
New Delhi-110 002

Environmental Biology

First Edition 1991

Reprint 1993, 1995
Reprint 1998

PRINTED IN INDIA

Published by Akashdeep Publishing House, New Delhi-110 002
and Printed at Mehra Offset Press, Delhi.

Preface

In many universities, environmental biology has been introduced as a paper at the undergraduate as well as the post-graduate level. But there are very few books on this subject. Therefore, an attempt has been made to write the book on the said subject. The order of the topics covered in this book may not be able to satisfy every one. However, the order of the topics covered in this book can do some justification for students. Each topic covered in this book is self-sufficient in itself and is well-explained with numerous figures. Further an attempt has been made to describe each topic in the light of modern developments in a simple language and elegant style.

This book will meet the requirements of a wide range of readers who are involved directly or indirectly with environmental biology.

We would like to thank several of my immediate colleagues and friends for making useful suggestions during the preparation of manuscript.

Suggestions or criticism towards the further improvement of the book shall be gratefully acknowledged.

The Editors

Contents

Contents

1

Environmental Biology

Introduction

Due to the advancement in the knowledge of the environment, there occurs an explosive increase in human population, scarcity of space, food broblems, threatening depletion of natural resources and socio-economic problems. In order to meet the challenge of these problems, one requires extensive and exhaustive understanding of the environment especially in relation to human survival and benefit. This all comes under the subject called *environmental biology*.

It is of interest to note that environmental biology is not a new discipline but is simply an extension of ecological approach which is mainly concerned with the study of the environment in relation to the welfare of man and his comfort. The reason for this is probably that man is regarded as the most evolved and an intelligent component of the environment.

Environmental biology is not a single disciplinary science but a multidisciplinary science which encompasses many vastly different fields such as geography, climatology, pedology microbiology biochemistry, physiology, biophysics, biometry, sociology, economics, etc.

A basic understanding of the fundamental concepts of environmental biology is not only essential for all biologists and environmentalist but also for non-biologists and non-environmentalists who are engaged in environmental science and related fields.

The Biosphere—Ecosystems

The biosphere, defined as the entire area of the earth that is inhabited by organisms, extends from the ground sector (the pedosphere) over the water zone (the hydrosphere) into the layer of atmosphere that envelops the earth. The functional units of the biosphere are the ecosystems in the form of forests, savannas, lakes, seas, etc., which are in a constant phase of development and are consequently capable, to a certain extent, of self-regulation. Ecosystems are operating complexes for life communities (biocenoses) and nonbiological environmental factors.

The composition and classification of life communities in ecosystems are determined, nter alia, by climatic factors such as the precipitation level, solar radiation, and the temperature and movement of the atmosphere. Thus the spread of holly (Ilex aquifolium) is due to climatic factors. The boundary of its distribution coincides with a hypothetical line through the points at which a maximum temperature of over 0°C (32°F) is achieved on 345 days in the year. In the Kerguelen Archipelago in the Indian Ocean, a group of islands constantly wracked by severe storms, flying insects have become extinct from constant exposure to winds carrying them into the sea. The earlier types of flying insects have heen supplanted by wingiess varieties and insects with a restricted flight range.

Within the climatic limits to a plant's growth, the qua'ities of the soil naturally detemine the emergence and frequency of particular species. A distinction thus has to be made between lime-loving and lime-resistant plant, and between salt-loving and salt-resistant plants.

It is not true today to say that only abiotic factors affect life communities and their distribution; it is truer to say that biocenosis (see also later discussion) exercises an influenee on the environmental factors. The strength of the wind, for example, which is of great importance to evaporation, is much reduced in woodland areas, and in a thick forest drops to a value of practically nil. Forest land also influences local temperature conditions : in a wooded region the extremes of temperature compared with those occurring on open land are much more moderate. Heat from the sun's rays is almost wholly consumed at the crowns of trees. On the other hand, the cold air that arises as a result of thermal radiation from the outer surfaces of the leaves on the crowns of trees descends to trunk level and mixes with the warmer air present there. Consequently there is no cold air to flow down wooded slopes, which is a vegetation advantage on the downward side.

As regards their energy balance, ecosystems are open systems that absorb energy from the sun unilaterally. The natural matter cycles in an ecosystem are balanced, so that a dynamic equilibrium, a so-called balanced flow, sets in. The organisms are traversed by a constant flow of exchange material and energy in the form of nutrients. The balanced flow produces a quasi-stationary condition in relation to the chemical compounds involved in the exchange processes. This emerges from the fact that the concentration of both the gases, carbon dioxide and oxygen, involved in the photosynthesis of plants (see Part Six) shows no change over a long period.

The ecological balance is characterized by the fact that every change in the ecosystem automatically promotes a

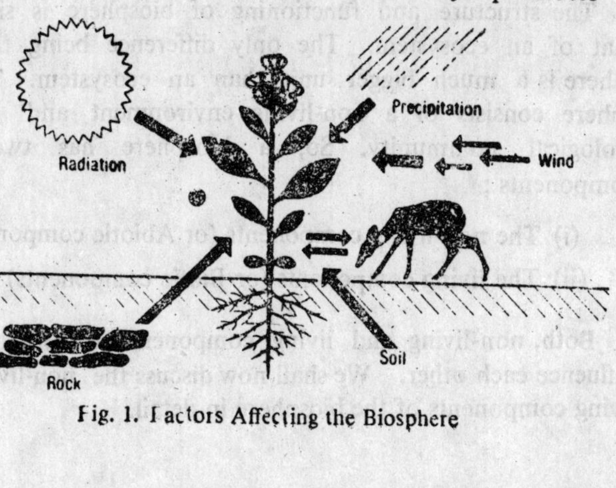

Fig. 1. Factors Affecting the Biosphere

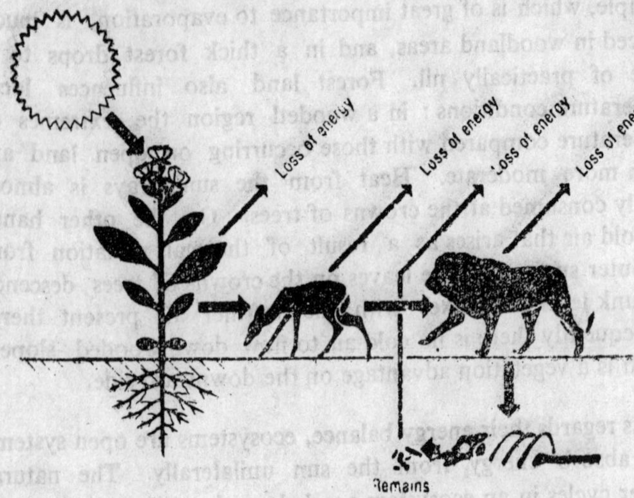

Fig. 2. Energy Flow in an Ecosystem

counterchange through an automatic control system, and this to a large extent reestablishes the previous situation. Thus in the case of voles and hares, pronounced population waves are distinguishable which are explained by the fact that as the population expands either the nutrient base becomes scarce or the vermin and parasites also increase. The system fluctuates as a rule between specific boundary values.

Components of the Biosphere

The structure and functioning of biosphere is similar to that of an ecosystem. The only difference being that biosphere is a much bigger unit than an ecosystem. The biosphere consists of a non-living environment and a living biological community. So, a biosphere has two main components :

(i) The non-living components (or Abiotic components)

(ii) The living components (or Biotic components)

Both, non-living and living components of the biosphere influence each other. We shall now discuss the non-living and living components of the biosphere in detail.

The Non-Living Components of Biosphere (Abiotic Components)

The non-living components of the biosphere consist of the physical environment like soil, water and air alongwith the inorganic substances like carbon dioxide, nitrogen, oxygen, water, phosphorus, sulphur, sodium, potassium, calcium and other elements present in them. The factors like sunlight, temperature, pressure and humidity are also a sort of non-living components of the biosphere and they constitute the climatic conditions. The climate in an area alongwith the nature of soil and water in that area determines (decides) the type of organisms which will be found in that part of the biosphere.

The Living Components of Biosphere (Biotic Components)

The living component of the biosphere consists of three organisms called : Producers, Consumers and Decomposers. The produces, are plants, consumers are animals whereas decomposers are microscopic non-green plants. So, in simple words we can say that the living component of the biosphere consists of plants and animals. That is, the "biota" or living organisms in the biosphere are plants and animals. It is on the basis of their mode of nutrition that all the living organisms of the biosphere have been divided into three groups called producers, consumers and decomposers. Producers are autotrophs; consumers are heterotrophs whereas decomposers are saprotrophs. We shall now discuss the producers, consumers and decomposers in detail.

1. Producers

Those organisms which produce food are called producers. Producers are the organisms which can prepare their own food from simple inorganic substances like carbon dioxide and water, in the presence of sunlight. The examples of producers are green plants including algae. The green plants synthesise their food during photosynthesis by taking raw materials from the earth and energy from the sun. The green plants produce carbohydrates by photosynthesis and also synthesise proteins and fats. Thus, the green plants are called producers in the living world. Producers are the autotrophic organisms (selffeeder organisms) in the biosphere upon which other organisms depend for food. It should be noted that all the plants cannot be included in the category of producers. Only green plants are producers. The non-green plants which cannot do photosynthesis to produce food are not included in the category of producers. We have just learnt that green plants produce food. Now, food contains energy. It means that the green plants or producers produce energy. An important question now arises : Do the green plants (which are called producers) really produce energy ? No, the green plants or producers do not produce energy in the absolute sense of the word. The green plants only convert sunlight energy into chemical energy of food. So, in terms of energy, the green plants are actually converters (of energy) of transducers.

2. Consumers

Those organisms which consume food (eat food) prepared by producers are called consumers. The consumers depend on producers for food, directly or indirectly. The consumers get their food by eating other organisms or their products. In most simple words, consumers are the organisms that eat other organisms. All the animals are consumers. Even the microscopic animal life of the water called protozoa are consumer organisms. The examples of common consumer organisms are man, goat, deer, fish, lion, cow, and buffalo, etc. The cow and buffalo eat green grass and other green fodder because green grass and other green plants are producers of food. The biomass of grass and plants supplies food and energy to these animals like cow and buffalo. It should be noted that the consumer organisms like animals cannot prepare food from simple inorganic substances through photosynthesis. The consumers need ready-made food for their survival which they get from producers (green plants), either direcctly or indiretly. If an animal eats grass or other green plants or their products itself we say that it gets the food from produers directly. For example, a goat gets the food from producers directly when it eats grass. On the other hand, if an animal eats the meat of another animal (which eats grass), then we say that it gets the food from producer indirectly. For example, a lion gets food by eating goat which in turn eats grass. So, in this case the lion gets its food indirectly from producer grass (through the goat). The animals which get their food by eating the producers (plants) directly are called primary consumers or first order consumers. The herbivores such as cattle, deer and goat, etc., which derive their food (or energy) directly from plants or producers are called primary consumers or first order consumers. The carnivorous animals like lion, tiger, snake, hawks, and frog, etc., which feed on herbivores (primary consumers) are called secondary consumers or second order consumers. The consumer organisms are also called heterotrophs. Since a man can eat plant materials as well as animals, he can be a primary consumer or a secondary consumer depending on the situation. A man is called an omnivore.

3. Decomposers

The non-green micro-organisms like some bacteria and fungi, which are incapable of producing their food, live on the dead and decaying (decomposing) plants and animal bodies, and are consumers of a special type, called decomposers. We can now say that : The micro-organisms which break down the complex, organic compounds present in dead organisms like dead plants and animals and their products like faeces, urine, etc., into simpler substances are called decomposers. The examples of decomposers are certain bacteria and fungi The bacteria which act as decomposers are called putrefying bacteria. The bacteria and fungi act as decomposers by the secretions of their body surfaces which decompose the organic matter present in dead plants and animals into simpler substances and liberate ammonia, carbon dioxide, etc. They absorb some of these simpler substances for their own maintenance and release the remaining into the soil, water and air to be used by the producers again. The decomposers are also known as micro-consumers or saprotorphs. **They are a kind of heterotrophic organisms.**

Importance of Decomposers

The decomposers help in decomposing the dead bodies of plants and animals, and hence act as cleaning agents of environment. The decomposers also help in putting back the various elements of which the dead plants and animals were made, back into the soil, air and water for re-use by the producers like crop plants. This maintains the fertility of soil, and the soil would continue to support crops again and again. For example, the decomposers like puterfying bacteria and fungi decompose the dead plants and animal bodies into ammonia (and other simpler substances). This ammonia is converted into nitrates by the nitrifying bacteria present in soil. These nitrates act as fertilizer in the soil and are again absorbed by the plants for their growth. Thus, it is only due to the presence of decomposers that the various nutrient elements which were initially taken by plants from the soil, air and water are returned to the soil air and water, after the death of plants and animals. If, however, there were no decomposers, then the dead bodies of plants and animals would keep lying as such and the elements of which plants and animal bodies are made, would never be returned to their original pools like soil, air and water. In that case, the cyclic process of life and death would be disrupted. This is because in the absence of decomposers the soil, air and water would not be replenished by elements from the bodies of dead organisms. All the nutrients present in soil, air and water would soon be exhausted and evolution of life would come to an end after that. Thus, the decomposer organisms help in recycling the material in the biosphere so that the process of life may go on and on like an unending chain.

Biome

The climatic conditions of an area like sunlight, temperature, rain fall, humidity and atmospheric pressure are important in determining the nature of plants and animals which can survive there. Due to the differences in the climates, the communities of plants, and animals very (change) over the surface of earth. However, same type of plants and animals appear together in similar climatic zones. The natural ecological groupings of plants and animals on the basis of climates are called biomes. In other words, all the ecosystems taken together in a given geographical area having the same type of climate is called a biome. A biome is a distinct large area of earth with relatively homogeneous flora and fauna (same type of plants and animals). The examples of biomes are : Desert; Grass-lands; Tundra; Tropical Forests; Temperate Forests; Coniferous Forests; Fresh-water, and Marine biome (or Ocean biome). The first six of these biomes are terrestrial biomes (land biomes) whereas the last two are aquatic biomes (water biomes). A biome is actually a very large ecosystem having the same type of climate, same type of plants and same type of animals throughout. So, biomes are also termed as major ecosystems of the world. For example, in a desert biome, we have the same type of hot climate throughout. And the same variety of plants (if any) and animals are found in the whole of desert biome. Though a biome is a climationally uniform area, it can have many different ecosystems in it. For example, in a desert biome also we can find some ponds and lakes and even some grars. All the biomes of the world taken together form an extremely large and self-sustaining biological system called biosphere. Before we go further and discuss the biosphere in detail please answer the following question :

Herbivores

The animals which consume plants or eat plants only are called herbivores (herb=plant; and vores=consuming). Goat, deer and cow are herbivores because they eat herbs or plants. Thus, the plant-eating animals are called herbivorous animals. Herbivorous animals are also called "First order consumers" (I order consumers).

Carnivores

The animals which eat only meat (of other animals) are called carnivores. Thus, the meat-eating animals or flesh eating animals are called carnivorous animals. Lion is a carnivore or carnivorous animal because it only eats the meat of other animals like deer, sheep. goat, buffalo, etc. In fact, all the predators are carnivores. A carnivorous animal is called second order consumer (II order consumer). And a large carnivore or top carnivore which eats second order consumer is called a "third order consumer" (III order consumer), and so on.

Omnivores

The animals which eat both plant food as well as meat of animals are called omnivores. Thus, 'plant-eating' as well as 'animal-eating' animals are called omnivorous animals. Man is the best example of an omnivore or omnivorous animal because he plant food as well as meat of animals (like goat, chicken, fish, etc.).

Planktons

Planktons are very minute or microscopic organisms freely floating on the surface of water in a pond, lake, river or ocean. Planktons are of two types : Phytoplanktons and Zooplanktons. The microscopic aquatic plants freely floating on the surface of water are called phytoplanktons. The free-floating algae is an example of phytoplankton. The microscopic aquatic animals freely floating on water are called zooplanktons. The freely-floating protozoa are an example of zooplankton. A very, very small fish is also a zooplankton.

Population

The group of organisms of the same species that share some common attributes (common characteristics) and are somewhat isolated from other groups, is called population. In simple words, the individuals of any species at a place form its population. The population is a higher level of organisation in the living world. In each population, several individuals of a species of plants or animals are found. Before we go further and discuss the term "community", we should know the meaning of "biotic" and "abiotic". Biotic means relating to the living organisms or biological. Abiotic means relating to non-living or physical world.

Community

The various populations of organisms (plants and animals) in an area taken together, form a community (or biotic community). A community includes several populations of plants and animals of a particular locality which are associated with each other for their existence. The various populations of a community interact among themselves and also with their physical environment. For example, a man interacts with other persons in his family; he also interacts with persons in his neighbourhood, place of work, and at social gatherings. The man also interacts with plants and animals around him as well as with the physical environment like soil, water and air. From the above discussion we conclude that all the organisms living in a locality constitute a community.

Biogeochemical Cycles

A cycle is a series of changes which comes back to the starting point and which can be repeated. The term "biogeochemical" tells us that biological, geological and chemical factors are all involved. Let us discuss all this in detail. The minerals are absorbed by the plants from soil or water; carbon dioxide is absorbed from the air and used in photosynthesis. The main elements and compounds needed by the plants are : Carbon (C), Nitrogen (N), Oxygen (O); Sulphur (S); Phosphorus (P), and water (H_2O) [water gives mainly hydrogen element (H) to the plants]. These materials after entering the producer level (plants) in the ecosystem, get transferred to the consumer levels like herbivores and carnivores. When the consumer organisms die and decompose, the various elements present in their body are released into the soil, air and water. These elements are put back into circulation again through re-absorption by plants, and the process is repeaed again and again. Thus, the transfer or circulation of chem cal nutrients in the biosphere takes place through soil, water, air and living organisms The circulation of chemical nutrients like carbon, hydrogen, oxygen, nitrogen, phosphorus, calcium and water, etc., through the biological and physical world (or biosphere) is known as biogeochemical cycle. **In a biogeochemical cycle, the chemical nutrients or elements** flow in characteristic paths from non-living environment to the living organisms and back to the environment. It is these pathways which are known as biogeochemical cycles. There are four very important biogeochemical cycle in nature. These are; Carbon cycle; Nitrogen cycle, Oxygen cycle and Water cycle. We shall discuss these biogeochemical cycles in detai after a while.

Carbon and Oxygen Cycles

The carbon and oxygen cycles in the biosphere are closely linked. The two cycles together form a system that comprises microbial decomposition processes, the photosynthesis of green plants, and the respiratory chain of animal organisms.

The reservoir for oxygen and carbon dioxide i the biosphere s the atmosphere. This contains, by volume, 0.03 % carbon dioxide and 21% oxygen. The total volume of carbon dioxide in the atmosphere is estimated at about 2100 trillion kg.

At a rough estimate, the annual formation of carbon dioxide by the photosynthesis of land plants amounts to some 13-22 trillion kg. The carbon dioxide taken uy bo water plants should be well in excess of that for land plants. Under favorable conditions the annual absorption of carbon by sea algae is thus estimated at .60 grams per square meter of the sea's surface. As the carbon absorbed by photosynthesis comes almost exclusively from the atmosphere, on the face of it the carbon dioxide content of the atmosphere should be constantly declining. That this does not happen is due to the activities of microorganisms and the lower animals, which are regularly replenishing the carbon dioxide content of the atmosphere by the mineralization of organic matter and by respiration.

The oxygen present in the atmosphere derives almost exclusively from photosynthesis. This oxygen is required for the supply of energy needed to sustain life in the course of animal and vegetable respiration—namely, the decomposition of energy-bearing organic matter (sugar).

A large part of the carbon assimilated by plants serves to maintain their life functions and is immediately emitted again. The remainder is incorporated as a structural or reserve substance in the plant cells. By day the assimilation rate of plants is higher than the transpiration rate, so that an increase in the amount of organic matter is achieved. At night the photosynthesis comes to a halt, and by their transpiration the plants lose biomass—that is to say, they draw on reserve substances.

Animal organisms, which do not have the capacity to assimilate carbon also contribute by their respiration to the return of carbon to the atmosphere. A grown man, for instance, breathes out about 1 kg of carbon dioxide every 24 hours.

But all the animals and human beings living on the earth today would not be able to mineralize again (by respiration) the whole of the organic matter built up by the plants and so maintain the necessary atmospheric supply of carbon dioxide. This function is performed for the most part by microorganisms (bacteria, fungi and actinomycetes). These have the capacity to mineralize organic matter, especially such as is not utilizable by other organisms – *i.e.*, carcasses, excrement and fallen foliage—and so to produce enormous quantities of carbon dioxide. Thus the production of carbon dioxide per hectare of good-quality soil in which several billion bacteria are present in every cubic centimeter is estimated at 2-5 kg per hour. In woodland areas the respiration rate is even higher.

In earlier epochs of the earth's history carbon was absorbed by living organisms, so that it got into sediments and was withdrawn from the carbon cycle. Coal, natural gas and oil reserves all represent such established carbon deposits, from which carbon is similarly returned to the atmosphere in the form of carbon dioxide by means of heating appliances and heat engines.

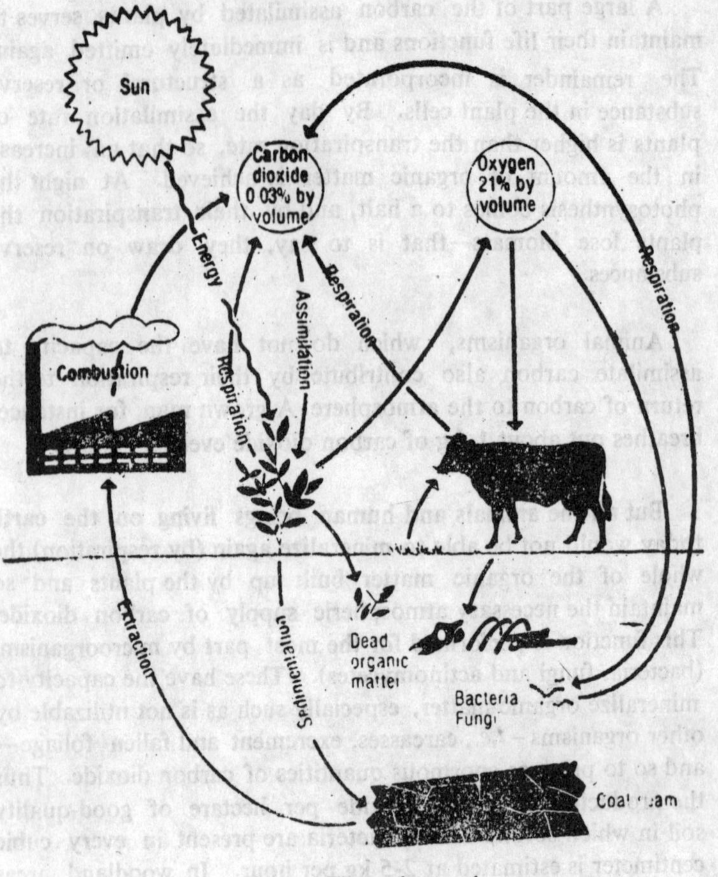

Fig. 3. The carbon cycle

Nitrogen Cycle

With a nitrogen content of 78%, the atmosphere is the most abundant nitrogen reserve on earth. Nitrogen is present there as N_2 (a nitrogen molecule containing two nitrogen atoms). In this form, nitrogen displays similar chemical properties to the inert gases. Under normal conditions prevailing on earth these hardly combine at all with other elements, being far from active. Consequently, with a few exceptions, the nitrogen in the atmosphere cannot be taken up and utilized by the organisms inhabiting the biosphere.

Nitrogen is one of the elements constituting the amino acids, which play a big part in all living organisms. The amino acids also represent the preliminary stage in proteins (albumins). The only organisms able to absorb nitrogen from the atmosphere are to be found among bacteria : nitrogen absorbers that live freely in the soil, and others that provide a symbiosis with plant roots. Both forms make the nitrogen available for autotrophic plants. When nitrogen absorbers die in the soil. the nitrogen present in the protein of these bacteria is broken down by other microorganisms into ammonia. For the optimum supply of nitrogen ro plants, the soil must also contain nitrates, which are likewise produced from ammonia in several intermediate stages, partly with the aid of soil-dwelling bacteria.

The plants absorb the nitrogen through their roots and incorporate it into the protein in their own systems. If plants go to ground or if parts of the plant drop off (as in the annual fall of foliage), they are also decomposed by microorganisms in the soil, and the nitrogen in thus passed on to the succeeding generation. Part of the nitrogen deposited in the soil by the nitrogenous bacteria is absorbed by rainwater and washed into the groundwater layer, whence it can be carried away into streams and rivers. The extent to which nitrogen is washed away depends largely on the volume of precipitation and also largely on the structure of the soil. In ground that is poor in humus and has a low clay content, the nitrogen is not retained (in the humus and soil particles) and is easily washed away. On sloping ground, clayey soils are exposed to erosion, as in these conditions precipitated water penetrates only with difficulty, and this also leads to a loss of the nitrogen adhering to the eroded soil particles.

Gaseous ammonia can also be released from the soil by microbial activity. This process is known as denitrification, because faced with a lack of oxygen the bacteria release oxygen from the nitrate in the soil. If the ammonia resulting from the decomposition of the protein is not converted into nitrate, gaseous ammonia can also be produced in certain circumstances.

The nitrogen absorbed by animals in conjunction with plant nutrients is partly converted into bodily protein and partly discharged in the urine or excrement. The nitrogen so emitted is also utilized microbially and restored to the plants. In the same way the body protein incorporated in the animal organism passes into the soil when the carcass of the dead animal decomposes.

Apart from the biological nitrogenous compounds formed by bacteria, nitrogen also passes into the ground upon discharge of electricity into the atmosphere in thunderstorms. Owing to the high level of energy that is released by such discharges, nitric oxide forms from the nitrogen and oxygen present in the atmos-

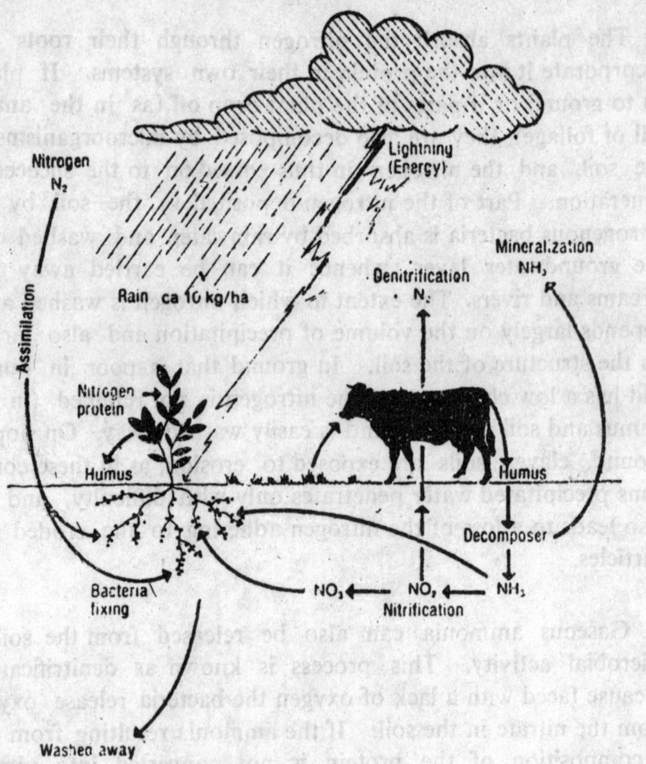

Fig. 4. The Nitrogen Cycle

phere, and this results in an increase in the nitrate content in the ground. This process is of importance, however, only in zones in which thunderstorms are frequent.

Phosphorus Cycle

Phosphorus is present in the biosphere almost exclusively in the form of phosphate. The elements calcium, iron and aluminum form phosphate salts, which are hardly soluble in water. The bulk of the phosphorus present on the earth is consequently integrated in rocks, soil or sediments and distributed fairly uniformly over the entire earth. In the course of the earth's history, deposits have formed at various points of the phosphorus-rich mineral apatite, and these have been used as natural reserves for the steadily rising phosphorus requirements of man.

Energy is required in connection with all the processes of life in the biosphere, and is converted by green plants from solar radiation into chemical energy. Of fundamental importance in the conversion process is adenosine triphosphate (ATP), a nucleotide consisting of one molecule of the base adenine, one molecule of the sugar ribose, and three molecules of phosphoric acid. As ATP is formed readily by the supply of energy from adenosine monophosphate (AMP) and adenosine diphosphate (ADP) and phosphoric acid, and decomposes again easily upon the release of the energy, it is a suitable compound for transferring energy for metabolic reactions or for storing energy released by decomposition reactions. In its combination with ATP, phosphorus is to be regarded as a key substance in life.

Phosphorus is obtained by plants out of the ground as a phosphate ion. Owing to the poor solubility of phosphorus in water, only a few phosphate ions are available in solution in the ground. The availability of phosphorus is heavily dependent on the soil acidity (pH value) and the form of phosphorus in the soil. In acid soils, iron and acid phosphates dissolve less easily as the decline in the pH value increases. If the ground is basic (alkaline), calcium phosphates are formed, the solubility of which declines as the pH value rises. Ground phosphates are available to plant life in large quantities when the soil reaction is neutral.

If the plants or parts of them die, the organic substance is returned to the soil and there decomposed by bacteria and fungi. In the process, part of the phosphorus incorporated in the plants will be absorbed by the microorganisms, and the rest will be released and pass into the ground. According to the soil acidity and the supply of iron, aluminum and calcium ions, the phosphorus will again be established in the form of iron, aluminum or calcium phosphate. If the plants are consumed by animals, the phosphorus contained in them will be ingested by the animal organism and ultimately be returned to the soil in excrement or with the carcass.

In the atmosphere, phosphorus is largely combined with aerosols. Its displacement in this form is of consequence only in dust, and sand-storms and volcanic eruptions. When the air turbulence subsides, the dust particles settle down again. Phosphorus normally gets into stream and river currents only through erosion of soil particles containing phosphorus. Phosphorus tends to dissolve in water, though to only a limited extent, and ro is absorbed by water plants. By far the greatest part is deposited with the eroded soil particles.

In natural ecosystems the subsequent delivery of phosphorus into the biosphere results primarily from the weathering of phosphorus-bearing rocks. Under certain conditions, in water meadows for instance, the supply of phosphorus in sludge (which is deposited through the periodic flooding of these regions) may play an important part in the provision of phosphorus to the vegetation, while the removal of phosphorus is significant only in extremely dry regions with little plant growth and a high incidence of wind erosion,

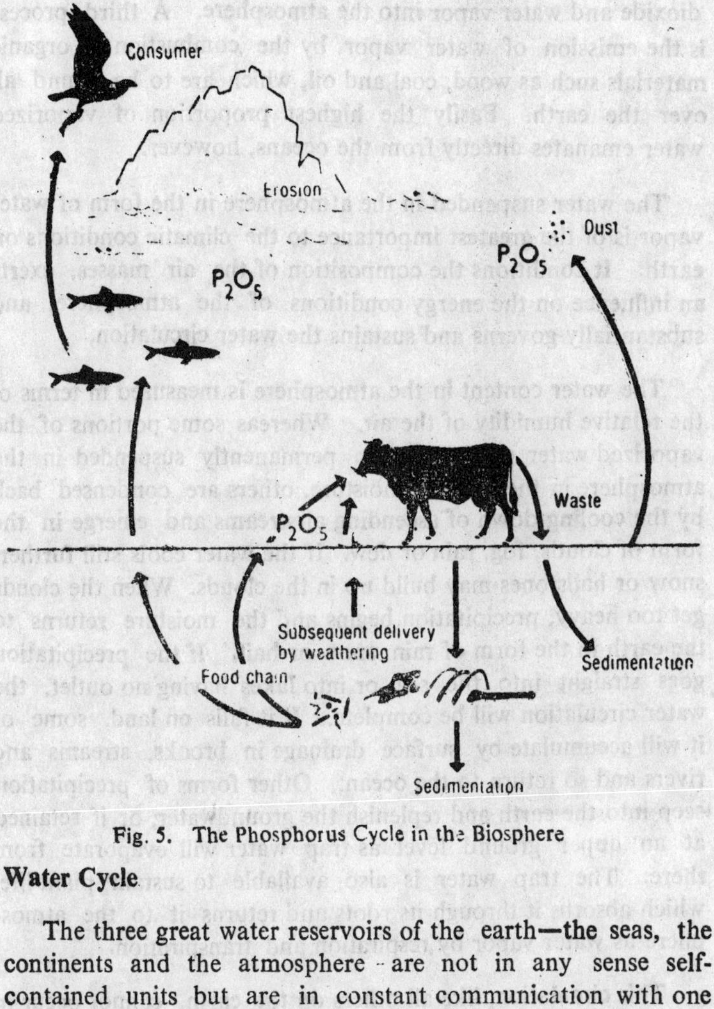

Fig. 5. The Phosphorus Cycle in the Biosphere

Water Cycle

The three great water reservoirs of the earth—the seas, the continents and the atmosphere - are not in any sense self-contained units but are in constant communication with one another. The relationships take the form of a circulation. Under the influence of solar radiation, water is constantly vaporized and emitted into the atmosphere in the form of water vapor. This process can occur in a great variety of ways. On the one hand, water evaporates directly from the surface of seas, lakes and rivers, from glaciers and snowfields, or straight out of the ground. On the other hand in the process of breathing, all living organisms, animals and plants alike, emit both carbon dioxide and water vapor into the atmosphere. A third process is the emission of water vapor by the combustion of organic materials such as wood, coal and oil, which are to be found all over the earth. Easily the highest proportion of vaporized water emanates directly from the oceans, however.

The water suspended in the atmosphere in the form of water vapor is of the greatest importance to the climatic conditions on earth. It conditions the composition of the air masses, exerts an influence on the energy conditions of the atmosphere, and substantially governs and sustains the water circulation.

The water content in the atmosphere is measured in terms of the relative humidity of the air. Whereas some portions of the vaporized water masses remain permanently suspended in the atmosphere in the form of moisture, others are condensed back by the cooling down of ascending airstreams and emerge in the form of clouds, fog, rain or dew. If the water cools still further, snow or hailstones may build up in the clouds. When the clouds get too heavy, precipitation begins and the moisture returns to the earth in the form of rain, snow or hail. If the precipitation goes straight into the sea or into lakes having no outlet, the water circulation will be complete. If it falls on land, some of it will accumulate by surface drainage in brooks, streams and rivers and so return to the ocean. Other forms of precipitation seep into the earth and replenish the groundwater, or if retained at an upper ground level as trap water will evaporate from there. The trap water is also available to sustain plant life, which absorbs it through its roots and returns it to the atmosphere as water vapor by respiration and transpiration.

This circulation, like all others on the earth, cannot occur in the absence of some motive force. The entire cycle is maintained by the regular inflow of solar energy. Of the total amount of energy absorbed by the earth, about one-third is utilized in maintaining the water circulation. The temperature, humidity and movement (wind) in the atmosphere together determine the level of evaporation.

Although the total volume of atmospheric moisture is very small in comparison with the total hydrosphere, owing to the circulation of the atmosphere an enormous amount of water movement takes place in the course of a year. To produce the annual precipitation on the earth of about 470,000 cubic kilometers, the water in the atmosphere (about 12,300 cubic kilometers) must be replaced 35 to 40 times a year.

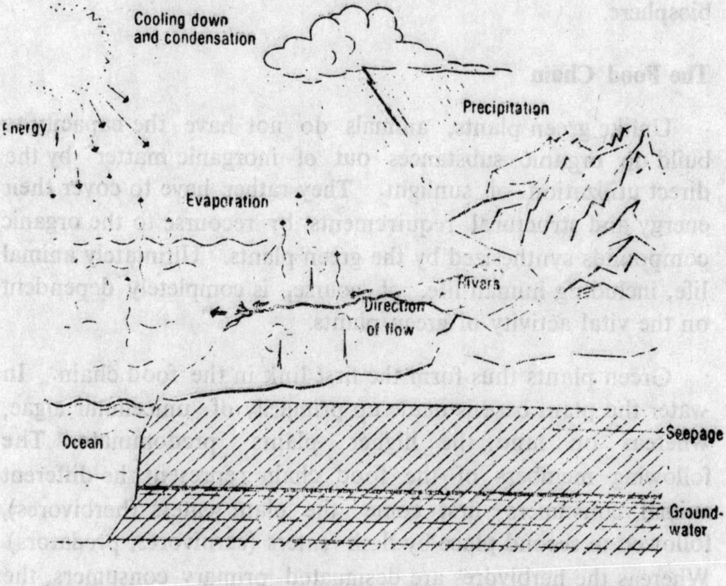

Fig. 6. Diagrammatic Presentation of Water Circulation in Nature

Disruption of Biogeochemical Cycles

The human activities like the modern agriculture and industries sometimes disrupt the normal working of biogeochemical cycles in nature. The various activities which can make the "cyclic" process of material circulation "acyclic" are :

(i) Excessive burning of fossil fuels.

(ii) Use of machines or vehicles having improper combustion mechanisms.

(iii) Over-use of chemicals like fertilizers and pesticides in agriculture.

(iv) Mining operations.

If the biogeochemical cycles of chemical elements or chemical materials get disrupted, then it will disturb the food chains in biosphere and lead to serious environmental hazards. So, in order to keep the environment healthy, we should practice modern agriculture and industrialisation with great care so as not to disturb the delicate balance of various systems in the biosphere.

The Food Chain

Unlike green plants, animals do not have the capacity to build up organic substances out of inorganic matter by the direct utilization of sunlight. They rather have to cover their energy and structural requirements by recourse to the organic compounds synthesized by the green plants. Ultimately animal life, including human life, of course, is completely dependent on the vital activity of green plants.

Green plants thus form the first link in the food chain. In water the plant base is made up primarily of unicellular algae, whereas on land the higher plants predominate. The following members of the food chain represent the different animal consumers : first come the plant eaters (herbivores), followed in second place by flesh eaters (carnivores, predators). Whereas the herbivores are designated primary consumers, the carnivores, which for their part consume herbivores or other carnivores, are classified as secondary, tertiary, etc., consumers according to their position in the food chain. Between decided plant eaters and exclusively flesh eaters come those that will eat anything (omnivores), which are classifiable as consumers at various points in the food chain and consume both animal and vegetable foodstuffs. Organisms with a wide food range are also described as polyphagous, as opposed to those that specialize in a particular diet, which are designated monophagous. An intermediate position is occupied by obligophagous creatures, which vary in the selection of their diet within certain limits.

The human being, who consumes anything that is palatable, is a typical example of a polyphagous animal. The oligophagous species include many insects that live on a small selection of food plants not related to one another. To the monophagous organisms belong many parasites that are only found on a particular type of host plant or animal and often only on certain of its organs. Head and body lice, for instance, concentrate exclusively on human beings.

The final link in the food chain is formed by decomposing organisms (demolishers and reducers). This group consists of bacteria, fungi and many ground-dwelling animals that live (as saprophytes or saprozoa) on decaying organic substances (excrement, detritus, carrion) - that is, on anything that is rejected or left over by members of the food chain already described. They live in humus and in the final analysis produce inorganic matter as required in return by the plants for their sustenance.

As a large part of the food in the food chain is converted for energy production, and a further portion is rejected as unutilizable, the weight of an organism increases at a rough estimate by only one tenth of the weight of food absorbed. From this it follows that when a human being consumes 10 pounds of pike meat, for instance, he will increase in weight by no more than 1 pound. In order to produce this 10 pounds of meat, the pike will, in turn, have to devour 100 pounds of carp, while the carp will need 1000 pounds of algae to enable them to increase their body weight accordingly.

In the food chain, the organisms consumed are normally smaller than those that live on them. Grain, for instance, is smaller than the mouse that consumes it, and the mouse in turn is smaller than a cat. The successive increases in size need not be pronounced, however, as in the sequence plant sap-plant louse-ladybug-songbird-bird of prey. On the other hand, the prey can be disproportionately large, like the victims of snakes, or it can be diminutive in size, as in the case of the giant bearded whale, which lives entirely on plankton.

Apart from various complications in detail, the general outcome was not a final reconciliation of the two concepts of

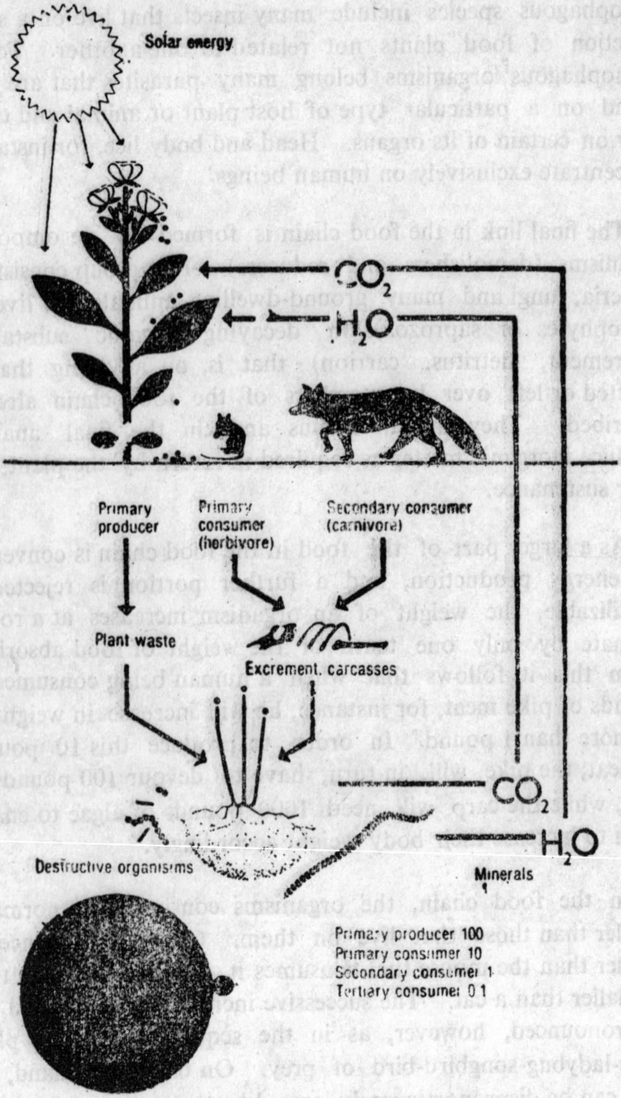

Decline in Biomass at Successive Stages in the Food Web

Fig. 7. The Individual Members of the Food Chain

reality so much as a realignment whereby each shared the full reality : the church the hereafter, and science the life in this world. Natural science, for its part, however, enjoyed the advantage of technical application. This led in the first place to arms production, and so to the possibility of the exercise of military power and finally to the development of industrial production. This development was characterized by liberalism, capitalism and imperialism. The influence and power of non-objective reality became ever weaker, and the notions of value and sense lost their influence on technical development. The old value concepts of truth, beauty and goodness, the ideals of measure, sense and order, paled into insignificance. Their place was taken by dynamic objects of technical development which may be set out as strivings toward preferences, the greatest possible volume and freedom of movement.

If scientific development had first restricted our perceptions of reality to the rational, as technical development advanced there would be further limitations. In fact, not everything is still rational, only what is scientifically and technically reproducible and feasible; ultimately a character of practical reality is still displayed only by what is economically interesting and viable. The scientific approach to reality is characterized by the segregation of individual problems and factors and by the causal-analytical investigation of mechanical functions. Everything that cannot be investigated in this context falls outside the scope of the exercise. Technical and economic science brings to the people an increasing sense of power. Science, which has made this fullness of power possible, assumes more and more the role of a kind of substitute for religion, from which by means of technical developments we promise ourselves a solution to all the evils of the world. This development is described by people as a belief in science and progress. The ultimate result is a world of science marked to an ever-increasing extent by specialized knowledge in which the layman has less and less say and the specialist gets increasingly involved.

If the old static value notions of standards, sense and order had led in their unilateral effects to lack of freedom, poverty and inflexibility as their negative counterparts, the dynamic maximum values of the technical period would lead, if they were realized unilaterally, to a striving for power, unlimited growth, and permanent change. This development has been described as an "abandonment of the middle course," but this completely overlooks the fact that the present one-sidedress is basically a consequence of the preceding static one-sidedress. This means that what was really necessary was nota return to the static notions, but a well-balanced equilibrium of the static and complementary dynamic ideas. In such an equilibrium, dynamic technical self-development had to be adjusted to people's ideas of sense and values again and related to the ecological limits of our possibilities. In our complicated scientific world, however, this was not possible without the help of a new science, reflecting fundamental aspects. Science was the seductive force making for the present industrial technology, and it must now lend its support to overcoming the difficulties that have arisen.

Functions of Food Chains

1. The food chains depicit (tell us) the structure of living components of the ecosystem or biosphere.
2. The food chains transfer energy and materials between various living components of an ecosystem or biosphere.

The transfer of energy and transfer of materials from one organism to another are the two basic functions performed by food chains which give dynamicity to an ecosystem or a biosphere. The transfer of energy and materials takes place in the predator and prey relationship in food chain. For example, when a predator like lion eats up its prey like deer, then the energy and material are tranferred from the deer to lion. The transfer of energy and materials takes place in different ecosystems such as grass-land, forest, park, gardens (which are terrestrial ecosystems) and in ponds, lakes, rivers, sea or even aquarium (which are aquatic ecosystems). The transfer of energy and materials does not depend on the size or shape of organism or the factors of the physical environment.

Before we go further, we should know the meaning of the term calorie. The calorie is a unit of measuring heat. One small calorie is the amount of heat necessary to raise the temperature of 1 gram of water by 1°C. This is very small unit, so the heat measuring unit used to reference to food is taken as kilocalorie which is equal to 1000 calories. When the kilocalorie unit is used for the heat energy of food it is called big calorie and written with a capital C as Calorie to distinguish it from small calorie. We can now say that big Calorie is the heat measuring unit used in foods (It is equal to 1000 small calories of the heat).

The Ten Per Cent Law

During the transfer of energy through successive tropic levels in an ecosystem, there is a loss of energy all along the path. No transfer of energy is 100 per cent. The studies of transfer of energy in different food chains in a large number of ecosystems has revealed in a uniform pattern of transfer of energy, which is given by 10 per cent law. The 10 per cent law which was given by Kindemann is the year 1942, is one of the most useful generalisations about the magnitude of loss in energy in food chains. According to ten per cent law only 10 per cent of the energy entering a particular trophic level of organisms is available for transfer to the next higher trophic level. All the energy transfers in a food chain follow the 10% law which in simple terms means that the energy available at each successive tropic level is 10 per cent of the previous level. Thus, there is a progressive decline (gradual reduction) in the amount of energy available as we go from producer level to the higher trophic levels of organisms. Let us take one example to understand the 10 per cent law more clearly.

Suppose 1000 joules of light energy is emitted by the sun which falls on the plants (called producers). We know that the plants convert only one per cent (1%) of the light energy falling on them into chemical energy of food. So, the energy which will be available in plant matter as food will be only 1% of 1000 joules, which comes to 10 joules. The remaining 1000 – 10 = 990 joules of light energy or solar energy which is not utilized by the plants is reflected back into the environment. This is shown in

Fig. 8. Please note that the ten per cent law will not apply at this stage. It will apply only in the transfer of energy in a food chain.

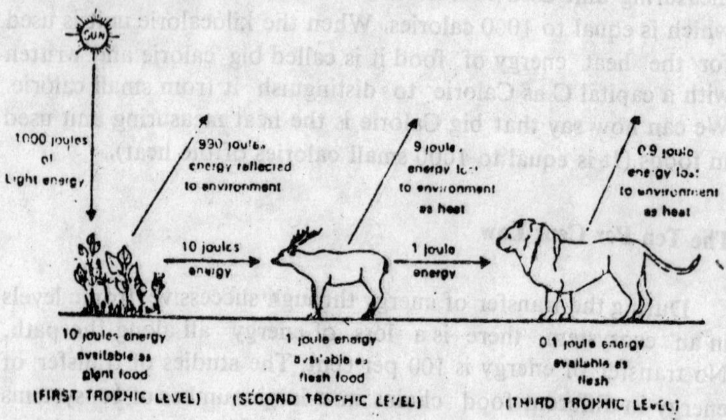

Fig. 8. Diagram to illustrate the 10 per cent law for the transfer of energy in a food chain

We shall now apply the 10 per cent law in the food chain : Plant → Herbivores → Carnivores (represented by plants, deer and lion in fig.). The plants or first trophic level has 10 joules of energy in it. Now according to ten per cent law, only 10% of 10 joules of energy (which is 1 joule) will be available for transfer at the next trophic level, so that the herbivore (deer) will get only 1 joule of energy at the second trophic level. The remaining 10 - 1 = 9 joules of energy will be lost to the env ronment as heat. Applying the ten per cent law again we find that 10% of the remaining 1 joule (which is 0.1 joule) will be transferred to third trophic l vel of carnivore (lion). The rest of it 1 - 0.1=0.9 joule will be lost to the environment. We shall now solve a problem based on the ten per cent law.

Food Web

The large number of food chains exist in a community of living organisms in an ecosystem such as a grass-land, a forest, a pond or a crop-field. Many of these food chains are interconnected by species (o ganisms) which occur in more than one food chain. The inter-connected food chains operating in an eco-system which establish a net-work of relationships between various species, is called a food web. In simple words, the net-work on a large number of food chains existing in an ecosystem is called a food web. The food web has many intercrosses and linkages among the various species (produces and consumers) present in it. This means that the various food chains in an ecosystem do not operate in isolation (or alone). They operate in the form of a net-work of food-chains called food-web. A food-web is shown below in Fig. 9.

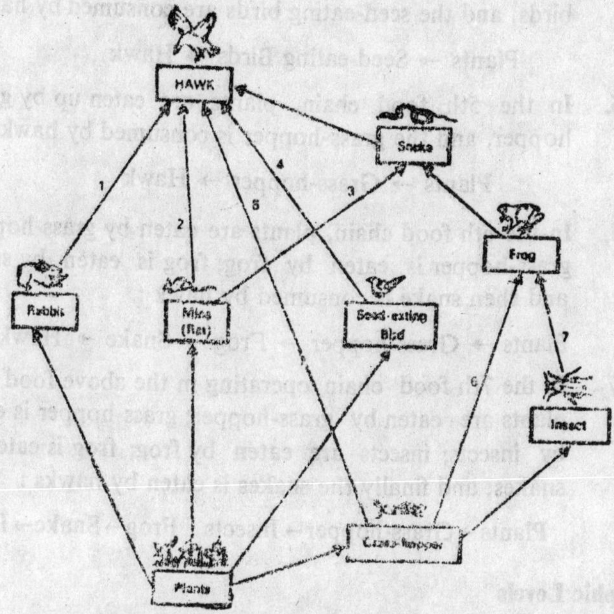

Fig. 9. Diagram to show a food-web

In this food-web, we can see a net-work of numerous pathways along which the food (or energy) flows within grassland community. This foodweb starts from the plants which is a producer and ends in top carnivore hawk (Baaz). There are as many as seven food chains operating in the food web shown above which have been marked 1, 2, 3, 4, 5, 6 and 7 (Fig. 9).

1. In the 1st food chain, plants are eaten by rabbit and then rabbit is eaten by hawk :

Plants → Rabbit → Hawk

2. In the 2nd food chain, plants are eaten by mice (or rats) and the mice are eaten by hawks :

   Plants → Mice → Hawk

3. In the 3rd food chain, plants are eaten by mice; mice are eaten by snakes; and then snakes are consumed by hawks :

   Plants → Mice → Snake → Hawk

4. In the 4th food chain, plants are eaten by seed-eating birds, and the seed-eating birds are consumed by hawks:

   Plants → Seed-eating Birds → Hawk

5. In the 5th food chain, plants are eaten up by grass-hopper, and the grass-hopper is consumed by hawks :

   Plants → Grass-hopper → Hawk

6. In the 6th food chain, plants are eaten by grass-hopper, grass-hopper is eaten by frog; frog is eaten by snake and then snake is consumed by hawk :

   Plants → Grass-hopper → Frog → Snake → Hawk

7. In the 7th food chain operating in the above food web, plants are eaten by grass-hopper; grass-hopper is eaten by insects; insects are eaten by frog; frog is eaten by snakes; and finally the snakes is eaten by hawks :

   Plants → Grass-hopper → Insects Frog → Snake → Hawk

Trophic Levels

A food chain represents the flow of food (or energy) in a given set of organisms or living beings. The various links (or steps) in a food chain at which the transfer of food or energy) takes place are called trophic levels. In fact, in food chain, each link representing an organism is known as a trophic level :

(i) The plants are producers and constitute the first trophic level.

(ii) Herbivores (which feed upon plants) constitute the second trophic level.

(iii) Carnivores (that feed upon herbivores) constitute the third trophic level.

(iv) Large Carnivores or Top Carnivores (which feed upon the small carnivores), constitute the fourth trophic level.

The various trophic levels in a food chain can be represented diagramatically as shown in Fig. 10. The simplest food chain that we have already studied is :

Grass → Deer → Lion

Now, this food chain involves three trophic levels. Grass (being producer) represents the Ist trophic level. Deer (being herbivore) represents the 2nd trophic level, and lion (being carnivore) represents the third trophic level. Here, deer is the first order consumer and lion is the second order consumer.

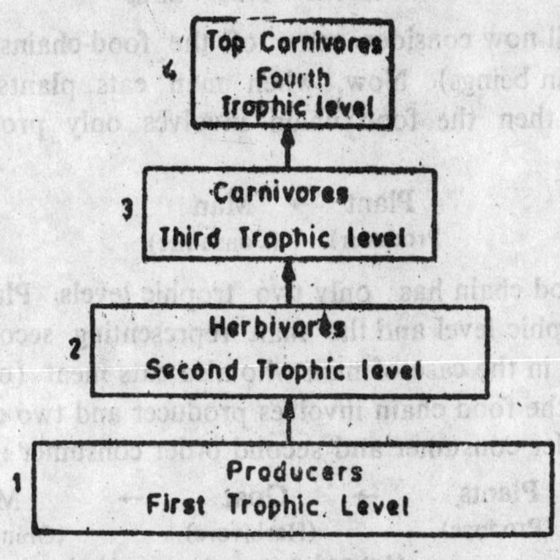

Fig. 10. Diagram to show the various trophic levels in a food chain

We have also studied another food chain operating in the grass-land, which is :

Grass → Insects → Frog → Birds

In this food chain, grass represents the I trophic level; insects represent the II trophic level; frog represents the III trophic level, whereas birds represent IV trophic level. This is shown more clearly in figure.

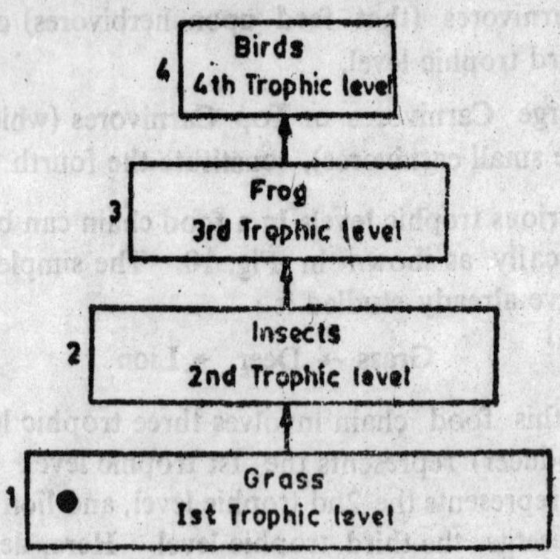

Fig. 11. Diagram to show trophic levels in food chain
Grass – Insects — Frog — Birds

We shall now consider some of the food-chains involving man (human beings). Now, when man eats plants (or plant products), then the food chain involves only producer and consumer :

$$\underset{\text{Producer)}}{\text{Plant}} \rightarrow \underset{\text{(Consumer)}}{\text{Man}}$$

This food chain has only two trophic levels. Plants being the first trophic level and the man representing second trophic level. But in the case of man who also eats meat (or animals like goat), the food chain involves producer and two consumers, the first order consumer and second order consumer :

Plants	→	Goat	→	Man
(Produce)		(Herbivore)		(Omnivore)
		(1st order consumer)		(2nd order consumer)

This food chain involving man has three trophic levels. Plants represent 1st trophic level, goat represents 2nd trophic level whereas man represents the 3rd trophic level.

The trophic levels in a food chain can also be represented by pyramid of numbers. Thus, if we count the number of species (or organisms) living at each trophic levels in a food chain, then we can represent the food chain by a pyramid of numbers. A pyramid of numbers showing the various trophic levels in the grassland food chain: Plants→Mice→Snakes—

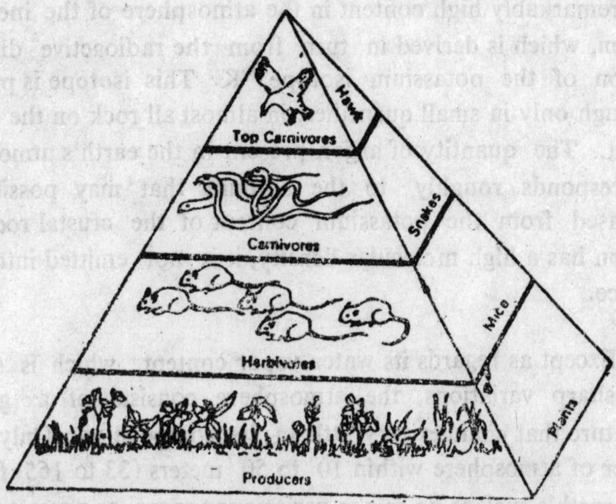

Fig. 12. Pyramid of numbers showing the various trophic levels in a food chain

Hawks is shown in Fig. 12. The base of this pyramid is formed by producers (plants) and the top of this pyramid is formed by the highest order consumers (or top carnivores). It should be noted that the same ecosystem may be supporting many different pyramids of organisms, each starting with plants at the base but ending in a different organism at the top.

The Atmosphere

The presence of a gaseous atmoshere around the earth may be due to a balance existing between the forces of gravity, which hold gases (in the same way as other masses) to the earth's surface, and the characteristic motion of molecules, which left to themselves would let the gases evaporate rapidly into outer space by diffusion. The molecular motion is greater the smaller the molecular and atomic masses and the higher the temperature. If the gases have low molecular masses, at constant temperatures they will attain higher velocities and so have a greater chance of escaping into outer space against the pull of gravity.

The atmosphere presumably owes its existence to exhalations (in the form of gases) from rocks. This thesis is supported by the remarkably high content in the atmosphere of the inert gas argon, which is derived in turn from the radioactive disintegration of the potassium isotope ^{40}K. This isotope is present, though only in small quantities, in almost all rock on the earth's crust. The quantity of argon present in the earth's atmosphere corresponds roughly to the amount that may possibly be released from the potassium content of the crustal rock. As argon has a high molecular density, it is not emitted into outer space.

Except as regards its water vapor content, which is subject to sharp variations, the atmosphere consists of a gaseous mixture that changes very little in its composition. Only in the layer of atmosphere within 10 to 50 meters (33 to 165 feet) of the earth's surface do minor variations appear, as a result mainly of the influence of organic life. The principal constituents of the atmosphere are the gases nitrogen and oxygen. The only other component present in significant proportions is the inert gas argan Apart from these there are a large number of so-called trace gases, which are present in such small quantities that they are measured not in the ordinary way by volume percentage but in ppm (parts per million). At a level of 1% to 4%, water vapour in the atmosphere ranks ahead of the inert gas argon. Deposite the comparatively low water element in the atmosphere, water vepor is still very important from the meteorological point of view since it is present as steam, liquid water, and ice.

Up to a height of about 100 kilometers (60 miles) the composition of atmospheric air is unchanged, apart from the rapid reduction in the water vapor content. Above 100 kilometers the atmosphere changes completely under the influence of ultraviolet radiation. The gases present in the lower atmosphere as polyatomic molecules break down at higher levels into their component parts.

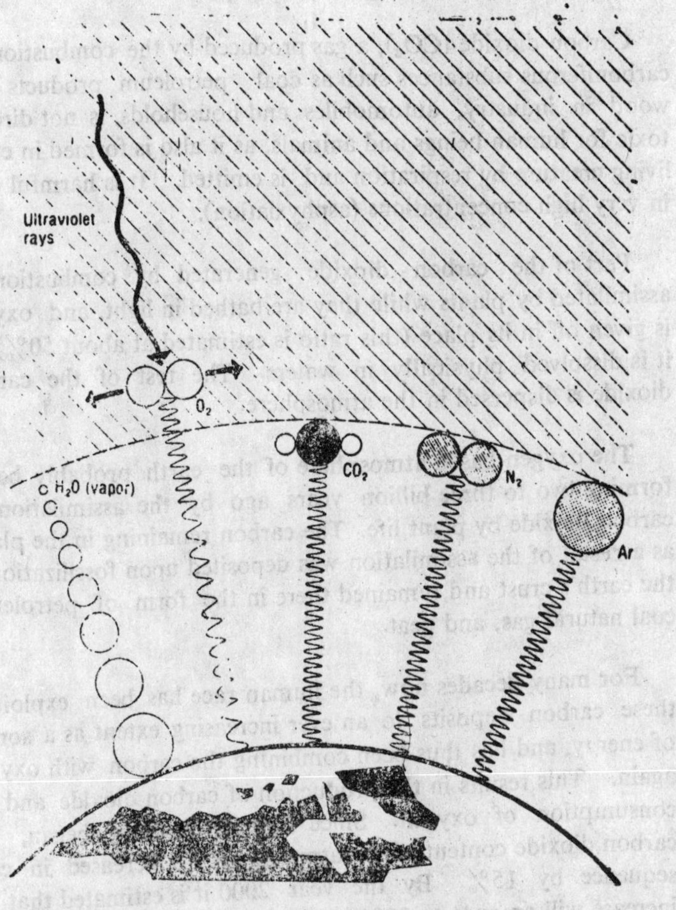

Fig. 13. Processes in the Atmosphere

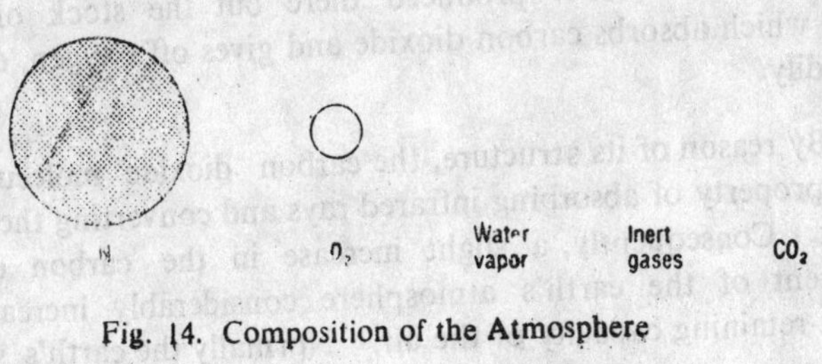

Fig. 14. Composition of the Atmosphere

Carbon Dioxide – Its Role in the Atmosphere

Carbon dioxide (CO_2), a gas produced by the combustion of carboniferous substances such as coal petroleum products and wood in industry, automobiles and households, is not directly toxic for human beings and animals, as it also is formed in every living creature by respiration and is emitted. It is harmful only in very high concentrations (asphyxiation).

Part of the carbon dioxide generated by combustion is assimilated by plants while they are bathed in light, and oxygen is given off in its place (this ratio is estimated at about 50%), or it is dissolved physically in waters. The rest of the carbon dioxide is dispersed in the atmosphere.

The oxygen-based atmosphere of the earth probably began forming two to three billion years ago by the assimilation of carbon dioxide by plant life. The carbon remaining in the plants as a result of the assimilation was deposited upon fossilization in the earth's crust and remained there in the form of petroleum, coal natural gas, and peat.

For many decades now, the human race has been exploiting these carbon deposits to an ever increasing extent as a soruce of energy, and has thus been combining the carbon with oxygen again. This results in the production of carbon dioxide and the consumption of oxygen. Since industrialization began, the carbon dioxide content of the atmosphere has increased in consequence by 15%. By the year 2000 it is estimated that the increase will amount to 30%. In industrial regions the carbon dioxide contained in the atmosphere is particularly high, because not only is more of it produced there but the stock of plant life, which absorbs carbon dioxide and gives off oxygen, declines steadily.

By reason of its structure, the carbon dioxide molecule has the property of absorbing infrared rays and converting them into heat. Consequently, a slight increase in the carbon dioxide content of the earth's atmosphere considerably increases the heat-retaining capacity of the air. Normally the earth's surface disseminates into the universe again a large part of the entering sunlight, in the form of infrared heat years. This prevents the temperature on the earth from constantly rising under the influence of solar radiation. But if the carbon dioxide content of the atmosphere increases, a larger part of this reflected infrared radiation will be absorbed into the atmosphere, and so will remain on the earth and thereby raise the temperature.

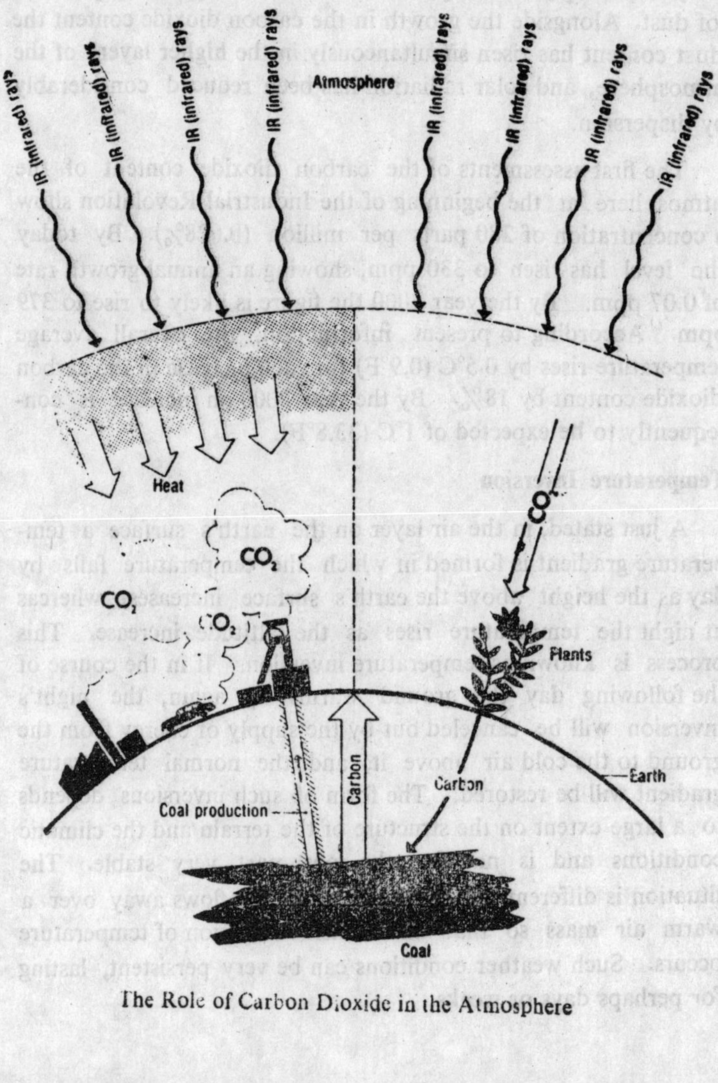

The Role of Carbon Dioxide in the Atmosphere

An increase in the temperature can change the climate of the earth considerably in the long run. A rise in average temperature throughout the world by 2°–3°C (35.6°–37.4°F) would very probably result in the melting of large quantities of polar ice, and so lead to a rise in the sea level and to the loss of large tracts of land. That this development has not yet occurred, despite the rising carbon dioxide content of the atmosphere, is due, surprisingly, to the pollution of the atmosphere by particles of dust. Alongside the growth in the carbon dioxide content the dust content has risen simultaneously in the higher layers of the atmosphere, and solar radiation has been reduced considerably by dispersion.

The first assessments of the carbon dioxide content of the atmosphere at the beginning of the Industrial Revolution show a concentration of 280 parts per million (0.028%). By today the level has risen to 330 ppm, showing an annual growth rate of 0.07 ppm. By the year 2000 the figure is likely to rise to 379 ppm. According to present information, the overall average temperature rises by 0.5°C (0.9°F) for each increase in the carbon dioxide content by 18%. By the year 2000 an increase is consequently to be expected of 1°C (33.8°F).

Temperature Inversion

A just stated, in the air layer on the earth's surface a temperature gradient is formed in which the temperature falls by day as the height above the earth's surface increases, whereas at night the temperature rises as the altitude increase. This process is known as temperature inversion. If in the course of the following day the ground warms up again, the night's inversion will be canceled out by the supply of energy from the ground to the cold air above it, and the normal temperature gradient will be restored. The form of such inversions depends to a large extent on the structure of the terrain and the climatic conditions and is not for the most part very stable. The situation is different when a layer of cold air flows away over a warm air mass so that an extensive inversion of temperature occurs. Such weather conditions can be very persistent, lasting for perhaps days or weeks.

The significance of the inversion for the exchange of gas in the air layer near the ground lies in the difference in specific gravity of colder and warmer gases in the same combination. Cold air heated up by the ground rises and cools. If it reaches the layer of warm air, because of its lower specific gravity any further rise of the cold air as it warms up is now checked. Thus, above the boundary layer no exchange of gas can occur.

Such thresholds create special problems for the layer of air near the ground in industrial agglomerations. The low-lying cold air is consumed in combustion processes in households, transport and industry, and the end product is further contaminated by combustion gases and dust. If no boundary layer existed, these noxious elements would be carried away into the atmosphere by normal dispersal processes and thereby dispersed to such an extent that there would be no immediate danger of injury to people, animals or plants. If an inversion occurs at a lower level (30 – 100 meters of 98-328 fit) above an industrial and housing area, the combustion products emitted are unable to penetrate the boundary layer and are retained in the lower layer of cold air. The concentrations of noxious gases and dust then increase rapidly and cause considerable damage to the biosphere (see Part Three).

In unfavorable valley regions and hollows the nightly inversion can also make itself felt in this way. That the inversion situation in valley basins is very stable is clear from the often immobile fog banks. Columns of smoke from chimneys stop at the boundary layers as if cut off; the smoke itself gathers below the boundary layer and forms great billows. If the chimney rises above the boundary layer the smoke can disperse freely. If the smoke is held down by high atmospheric moisture it will spread right across the boundary layer without penetrating it.

The frequency of such inversion situations depends on the season of the year. In summer, because of sustained and powerful solar radiation, there is very little inclination to inversion. As the day shorten, the excess of solar radiation over thermal radiation from the earth does not necessarily coincide with sunrise or sunset; and the thermal radiation map exceed solar radiation for several hours before sunset and several hours after sunrise. Solar radiation on the short winter days, thus, hardly has any effect even in fine weather, so that the frequency and duration of the inversion situation are greater in winter.

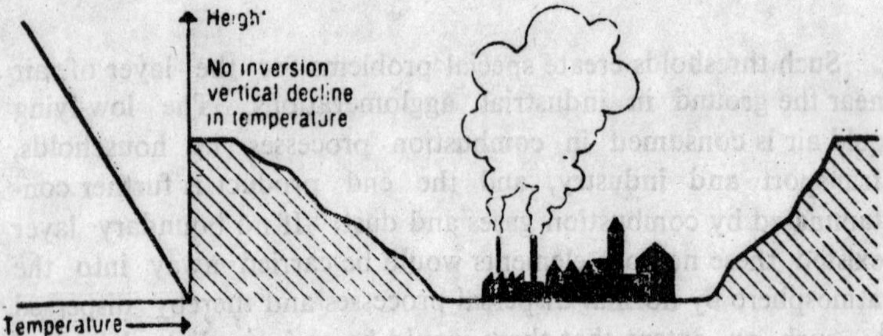

Fig. 15. Weather Conditions Without Inversion

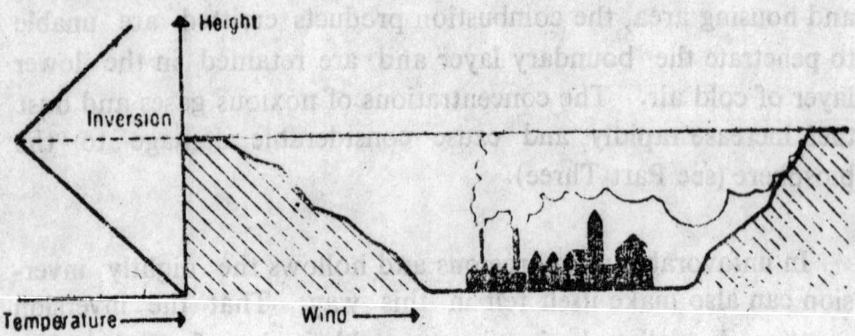

Fig. 16. Inversion Weather Conditions

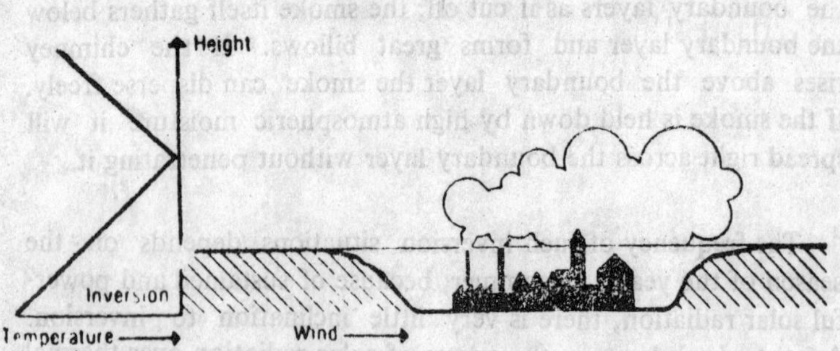

Fig. 17. Inversion Weather Conditions by Night

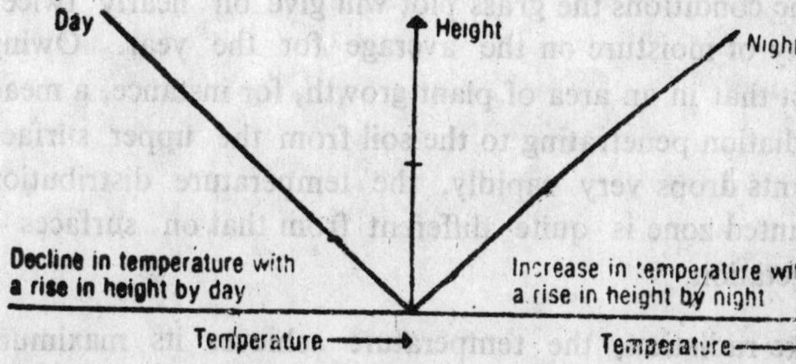

Fig. 18. Diagrammatic Temperature Movement in the Atmosphere by day and by Night

Temperature in the Plant Canopy

Temperature conditions similar to those for bare earth also apply to open plant communities (*e.g.*, young plantations). In extreme cases (southern slopes), when the wind is still and solar radiation on the earth's surface is at its peak, temperatures of 70°C may be attained within the canopy of leaves; that is, 20°—30°C (68°F – 86°F) above the temperature of the air. Temperatures of this order damage the young seedlings at the base of the stalks, as these will not yet have developed any final heat-insulating tissue. Sometimes this results in the withering and death of the plants. Most seriously affected by this are the shoots of trees. Cultivated plants that come out mainly in the spring and autumn are rarely exposed to such extremes of temperature.

If there is a close plant canopy covering the ground, the maximum temperature likely to be reached is reduced considerably. The active surface is no longer the surface of the topsoil but the tops of the plants exposed to the sun's rays. The temperature will now be reduced by the heavy transpiration of the plants. In open communities this is not possible, as evaporation stops at once as soon as the thin layer of soil dries out, whereas on ground covered with foliage the whole of the area occupied by the roots of vegetation gives off water to the atmosphere through the plants. In a comparison between a grass plot and bare sand, under the same temperate climatic conditions the grass plot will give off nearly twice the quantity of moisture on the average for the year. Owing to the fact that in an area of plant growth, for instance, a meadow, the radiation penetrating to the soil from the upper surface of the plants drops very rapidly, the temperature distribution in the planted zone is quite different from that on surfaces with no vegetation.

Like radiation, the temperature achieves its maximum at the topmost vegetation level, and owing to the poor heat-conducting capacity of the plant parts is not imported to the rootstock. As the height of the vegetation level increases, the maximum temperature changes. The leaf position in the vegetation is also an important factor for the level and the determination of maximum temperature. In a meadow in which the main mass of the leaves stands vertically, the rise in temperature (owing to the greater penetration range of the radiation) does not occur at one particular level but permeates the entire foliage space, so that extreme values are not attained. If the leaves in the vegetation are horizontal, the active surface of the foliage will draw together at the upper level so that the maximum temperature will be clearly determined at that point.

On clear nights the external layer of the foliage will be the radiating level (as in the case of bare ground), so that the minimum temperature will be established here, while the temperature in the vegetation will tend to rise near the ground. As a result, the temperature variations between day and night will be less marked in vegetation than in the case of bare ground. Consequently the average temperatures will be somewhat lower.

The temperature distribution in wooded areas will be determined on windless days by the distinction between an extreme crown-level climate on the active surface of the forest and a moderate trunk-level climate that is very balanced. The temperature conditions in woodland areas will also be influenced by the density of the forest. In light foliage, in which the sun's rays penetrate to the ground, there will be a second temperature maximum at that point. In dense foliage the ground temperature will correspond to that at the trunk level.

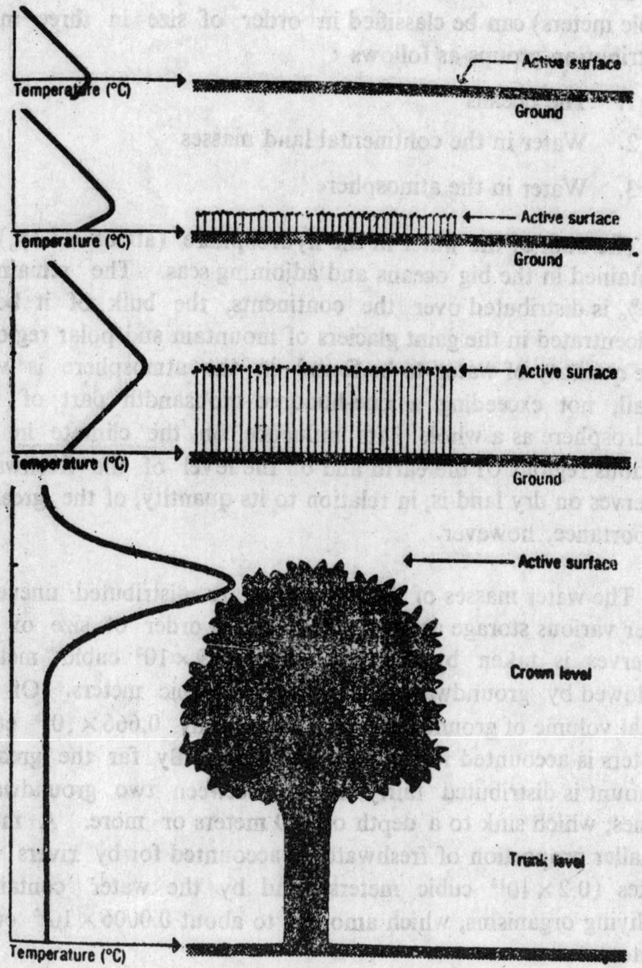

Temperature at the Vegetation Level

The Hydrosphere

Biological processes can continue on the earth only as long as water is available. For every living system the presence or provision of water is consequently of cardinal importance. The total volume of water reserves available on the earth today is known as the hydrosphere. Throughout the period of human evolution, its volume can be regarded as unchanged.

The water present in the hydrosphere (totaling 1.4×10^{18} cubic meters) can be classified in order of size in three main distribution groups as follows :

1. The oceans
2. Water in the continental land masses
3. Water in the atmosphere

The bulk of the water in the hydrosphere (about 97.3%) is contained in the big oceans and adjoining seas. The remaining 2.7% is distributed over the continents, the bulk of it being concentrated in the gaint glaciers of mountain and polar regions. The quantity of water to be found in the atmosphere is very small, not exceeding a one-hundred-thousandth part of the hydrosphere as a whole. Its influence on the climate in the various regions of the earth and on the level of the freshwater reserves on dry land is, in relation to its quantity, of the greatest importance, however.

The water masses of the continents are distributed unevenly over various storage areas. First place in order of size of the reserves is taken by the glaciers at 2.9×10^{3} cubic meters, followed by groundwater at 8.4×10^{15} cubic meters. Of the total volume of groundwater reserves, only 0.666×10^{15} cubic meters is accounted for by surface water. By far the greatest amount is distributed fairly evenly between two groundwater zones, which sink to a depth of 800 meters or more. A much smaller proportion of freshwater is accounted for by rivers and lakes (0.2×10^{15} cubic meters) and by the water contained in living organisms, which amounts to about 0.0006×10^{15} cubic meters.

Besides the continental freshwater tables, in the oceans there are two big freshwater reserves—namely, the water masses located in the polar ice of the Arctic and Antarctic, which together account for no less than 1.8% of the total hydrosphere, and so are relatively important.

This distribution of the water in ths hydrosphere among the various sectors has not remained unchanged throughout the course of history. Over the past two million years the ice caps of the polar regions have thawed out periodically and then

Hydrosphere as a whole	1.4×10^9 cu km	
Oceans	1.3×10^9 cu km	(97.3%)
Glaciers and polar ice	2.9×10^3 cu km	
Groundwater	8.4×10^6 cu km	
Lakes and rivers	0.2×10^6 cu km	(2.7%)
Atmosphere	1.3×10^4 cu km	
Biosphere	0.6×10^3 cu km	

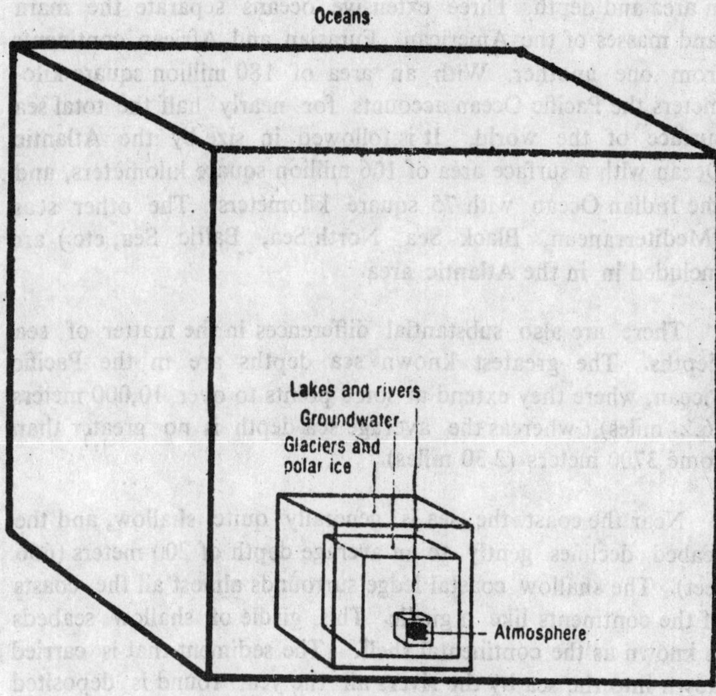

Distribution of water in the Hydrosphere

frozen again. If these ice caps were to thaw out completely today, the sea level would rise by over 60 meters, and the sea would, inundate large parts of the continents. In the periods of the greatest icing-up of the seas, on the other hand, the water level of the oceans has fallen by up to 140 meters. Such changes in the distribution of the water masses of the hydrosphere over the earth have had serious consequences, not only for these but also for other sectors, in the form, for instance, of droughts and flood disasters The lithosphere can also be changed by the emergence of heavy ice loads exerting pressure on the underlying rock

The Sea and the Seabed

The oceans cover nearly 71% of the earth's surface, or about 361 million square kilometers. This enormous volume of water is not spread evenly over the earth's surface, but varies sharply in area and depth. Three extensive oceans separate the main land masses of the American, Eurasian and African continents from one another. With an area of 180 million square kilometers the Pacific Ocean accounts for nearly half the total sea surface of the world. It is followed in size by the Atlantic Ocean with a surface area of 106 million square kilometers, and the Indian Occan with 75 square kilometers. The other seas (Mediterranean, Black Sea, North Sea, Baltic Sea, etc.) are included in in the Atlantic area.

There are also substantial differences in the matter of sea depths. The greatest known sea depths are in the Pacific Ocean, where they extend at some points to over 10,000 meters (6.2 miles), whereas the average sea depth is no greater than some 3700 meters (2.30 miles).

Near the coast the sea is generally quite shallow, and the seabed declines gently to an average depth of 200 meters (656 feet). The shallow coastal ledge surrounds almost all the coasts of the continents like a girdle. This girdle of shallow seabeds is known as the continental shelf. The sediment that is carried down into the sea by the rivers all the year round is deposited primarily in this area. The continental shelf is bordered by the continental margin, from which the continental slope drops away, From this point the seabed falls sharply to greater depths. This often provides gradients of 1 :15 and angles of gradient of over 45°. Occasionally a vertical face is found in this region. At about 3000 to 6000 meters (1.8 to 3.7 miles) from shore the continental slopes give way gradually to the deep-sea bed. The decline to the deep-sea bed seldom occurs in the continental slope in a uniform manner, however; it is often punctuated by hollows and ravines. The deep-sea bed proper is uniform and even over wide areas. But in certain sectors the seabed is intersected by still deeper troughs, the deep-sea trenches. At these points it may drop to a depth of 7000 meters (4.3 miles) or more. Most of these deep-sea trenches are to be found on the fringe of the Pacific adjoining Asia.

In other sea regions, on the other hand, the seabed rises to form big plateaus, which cover wide areas of the oceans. These plateaus, or submarine ridges, can be likened to the mountains on the continents, though they are much more sharply delineated. The biggest known sea obstacle is the Mid-Atlantic Ridge, which stretches in the shape of an S from north to south midway across the Atlantic Ocean and roughly follows the coastline of the American continent. Such submarine rides and plateus play a very important part in setting thepattern of ocean currents and the exchange of waters between the individual regions of the sea.

Ecological Pyramids

The food chains involving various living organisms or species in a community can be represented graphically by drawing pyramids which are known as ecological pyramids. Thus, ecological pyramids is another way of representing the energy transfer taking place in a food chain in a community. As we have already discussed the ecological pyramids also tell us the trophic levels or trophic structure in a community of organisms. The ecological pyramids depict the food chain graphically.

Three types of ecological pyramids can be constructed for a community of organisms in an ecosystem. These are ;

(i) Pyramid of numbers

(ii) Pyramid of mass (or biomass), and

(iii) Pyramid of energy.

The ecological pyramid constructed on the basis of number of organisms (or species) at each trophic level in a community is called pyramid of numbers. For example, we can count the number of plants, the number of mice, the number of snakes and the number of hawks in a given area or ecosystem, and construct the pyramid of numbers. Such a pyramid of numbers for the plants, mice snakes and hawks in a community is shown in Fig. 12 on page 32. In figure 12, the base of pyramid is very big and represents a large number of plants. We find that as we go up in the pyramide and the number of trophic level increases, the number of living organisms in it continuously decreases. Thus, the number of plants is very large, the number of mice is less, the number of snakes is still lesser, whereas the number of hawks is the least. The ecological pyramids constructed on the basis of mass (or biomass) of all the organisms at each trophic level is called pyramid of mass. A pyramid of mass is shown in figure. It is also possible to construct a pyramid of energy by estimating the energy contained in all the organisms of each trophic level to have a graphic representation of energy transfer at each level. The ecological pyramids show various trophic levels in the ecosystem. The base of an ecological pyramid is very big and represents the producer level like plants (or gross). As we go up, the pyramid goes on becoming thinner and thinner, indicating the successive trophic level in the food chain. Above the producers are herbivores and above herbivores are carnivores. The top carnivores or large carnivores from the top portion of the ecological pyramid. Let us take one example to understand the concept of ecological pyramids more clearly.

A study made in a grass-land areas showed that it contained plants, mice, snakes and hawks. The plants were eaten by mice; the mice were eaten by snakes and finally the snakes were eaten up by hawks. So, the following food chain existed in the grass land :

Plants → Mice → Snakes → Hawks

Further study showed that when 1000 kilograms of plants were consumed then 100 kilograms of mice were supported.

These mice became food for 10 kilograms of snakes which were themselves eaten up by 1 kilogram hawks. A pyramid of mass constructed from this data is shown in Fig 19. This pyramide of mass tells us that an enormous mass of plants is required to support a relatively much smaller mass of hawks; each stage in the pyramid has far less mass than the preceding one; and an imbalance in any part of the pyramid necessitates the adjustment in the rest of it. The broad base of pyramid is always composed of photosynthetic green plants (or grass), with successively higher trophic levels consisting of progressively larger but fewer animals.

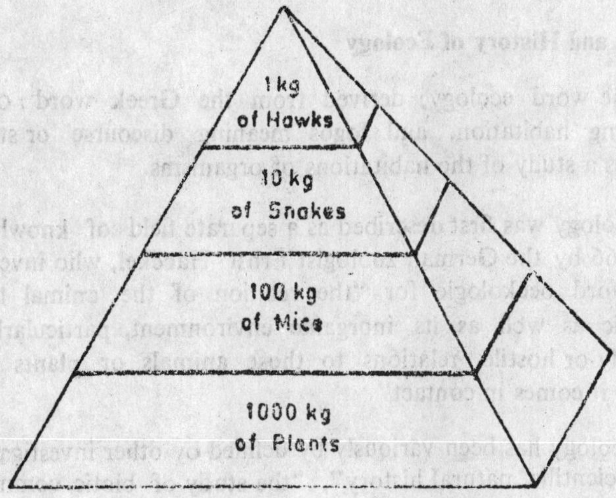

Fig. 19. A pyramid of mass : Graphic representation of principle of food chain

2

Concepts of Ecology

Scope and History of Ecology

The word ecology, derived from the Greek word : oikos meaning habitation, and logos meaning discourse or study, implies a study of the habitations of organisms.

Ecology was first described as a separate field of knowledge in 1866 by the German zoologist Ernst Haeckel, who invented the word oeckologie for "the relation of the animal to its organic as well as its inorganic environment, particularly its friendly or hostile relations to those animals or plants with which it comes in contact".

Ecology has been variously by defined by other investigators, as "scientific natural history", "the study of biotic communities", or "the science of community population"; probably the most comprehensive definition is the simple one most often given; a study of animal and plants in their relations to each other and to their environment.

Objectives

Ecology is a distinct science because it is a body of knowledge not similarly organized in any other division of biology; because it uses a special set of techniques and procedures; and because it has a unique point of view. The essence of this science is a comprehensive understanding of the import of these phenomena ;

1. The local and geographic distribution and abundance of organisms (habitat, niche, community, biogeography).
2. Temporal changes in the occurrence, abundance, and activities of organisms (seasonal, annual, successional, geological).
3. The interrelations between, organisms in populations and communities (population ecology).
4. The structural adaptations and functional adjustments of organisms to their physical environment (physiological ecology).
5. The behaviour of organisms under natural conditions (ethology).
6. The evolutionary development of all these interrelations (evolutionary ecology).
7. The biological productivity of natural and how this may best serve mankind (ecosystem ecology).
8. The development of mathematical models to relate interaction of parameters and predict effects (systems analysis).

Relation to Other Sciences

Ecology is one of their main divisions of biology, the other two being morphology and physiology. The emphasis in morphology is on how organisms are made, in physiology on how they function, and in ecology, on how they live. These divisions overlap broadly. To appreciate fully the structure of an organ, one needs to know how it functions, and the way it functions is clearly related to environmetal conditions. The morphologist is concerned with problems of anatomy, histology, cytology, embryology, evolution, and genetics; the physiologist, with interpreting functions in terms of chemistry, physics and mathematics; and the ecologist, with distribution, behaviour, populations and communities in relation to the environment (ecosystems). The evolution of adaptation and of species is mutual interest to the ecologist and to the geneticist; biometeorology is a connecting like between ecology and physiology; and system analysis interrelates ecology and mathematics. All areas, in the final analysis, are simply different approaches to an understanding of the meaning of life.

Subdivisions of Ecology

Ecology may be studied with particular reference to animals or to plants, hence animal ecology and plant ecology. Animal ecology, however, cannot be adequately understood except against a considerable background of plant ecology. What animals and plants are given equal emphasis, the term bioecology is often used. Courses in plant ecology usually dismiss animals as but one of many factors in the environment. Synecology is the study of communities, and autecology the study of species. There is some confusion in these terms since Europeans commonly use "ecology" in narrower sense meaning the environmental relations of organisms or of communities. The broader study of communities, including species interrelations and community structure and function as well as environmental relations (synecology), is generally termed "biocenology" or "biosociology" by Europeans.

We shall survey the fundamentals and basic facts of ecology as they relate to animals and have application to man. We will study community ecology, the local distribution of animals in various habitats, the recognition and composition of community units, and succession; ecosystem dynamics, the processes of soil formation, nutrient cycling, energy flow and productivity; population ecology, the manner of population growth structure, and regulation; evolutionary ecology, the problems of niche segregation and speciation; and geographic ecology, concerned with distribution, paleoecology, and biomes. We will be integrested throughout the text with how organisms respond and adjust physiologically to the physical factors of their environment, but a full study of physiological ecology must be left to another time and place. We will also be concerned throughout with systems ecology that is the possibility of translating ecological concepts into mathematical models, although we will not go deeply into the actual statistical manipulations.

Biotic Community

Structure and Dynamics. The concept of the biotic community is basic to an understanding of ecology. We will here be concerned only with laying a foundation of general principles. Details will come in later lesions, but for proper orientation we must know something about how the community is organized, how it functions, and how it may be recognized.

Internal Structure and Processes

Community and Ecosystem. A community, or biocenose, is an aggregate of organisms which form a distinct ecological unit. Such a unit may be defined in terms of flora or funa, or both. Community units may be very large, like the continent-wide coniferous forest, or very small like the community of invertebrates and fungi in a decaying log. The extent of a community is limited only by the requirement of a more or lessuniform species composition.

A different community occurs in each different habitat and environmental unit of larger size, and in fact the composition and character of the community is an excellent indicator of the type of environment that is present. Since plants and animals, bacteria and fungi, all occur together in the same habitat and have many interrelation they can scarcely be considered independently of each other. Together they make up the biotic commmunity.

Communities may be distinguished as major or minor. Major communities are those which, together with their habitats, form more or less complete and self-susttaining units or ecosystems, except for the indispensable input of solar energy. Minor communities, often called societies, are secondary aggregations within a major community and are not, therefore, completely independent units as far as circulation of energy is concerned.

The biotic community along with its habitat is called an ecosystem. The term "ecosystem" has been loosely applied to units of various sizes and characteristics but had best be limited to distinctive combinations of air, soil and water conditions with vegetative, animal, and microbic life.

Productivity. A characteristic of communities that has become of considerable importance in modern ecological search is productivity. The number of individuals of biomass in a community at any one time is the standing crop. At the beginning of the year or reproductive season the standing crop is usually small but as reproduction and growth take place there is an increase in the amount of organic matter making up the biomass of the community. The production of organic matter per unit of time and area is productivity. Productivity is thus a rate and is commonly indicated on a yearly basis, but it is also possible to measure monthly, weekly or daily production. Small standing crops may have a high productivity and large standing crops a low productivity, hence biomass alone does not give a measure of productivity. The largest standing crop which a habitat can support without deterioration, or the maximum number or biomass of animals that can survive the least favourable yet tolerable environmental conditions during a stated period of time, is the carrying capacity. Carrying capacity is determined not just by the amount of food available, but also by shelter, social tolerance, and other factors. A variety of methods are being used to measure productivity of different kinds of organisms and of different habitats. It is desirable to indicate productivity as accurately as possible in descriptions or analysis of community dynamics.

Species Diversity. Ecologists have long been aware that communities in different environments vary in the number of species that they contain, but only in recent years have quantitative indices been developed to show the relation between community structure not only (1) in number of species but also (2) in the relative number of individuals in each species.

Succession

Communities are in a more or less continual process of change (Clements 1916). These changess result in part form the reactions and coactions of the organisms themselves and in part from such external forces as changing physiography, changing climate, and organic evolution. The habitat is usually affected as well as the community, and as the habitat changes, new species invade it and become established, and old species disappear. These changes are especially noticeable in dominant species, since these species exert a controlling role over the composition and structure of the community as a whole. The replacement of one community or ecosystem by another is succession, and succession continues until a climax or final stage is reached.

Succession is a process. The series of steps or communities comporising a successional sequence leading to the climax is the sere. Seres are sometimes classified according to the predominant force that is bringing them about. These forces are biotic, climatic, physiography, and geologic and their resultant seres are commonly called bioseres, cliseres, eoseres and geoseres.

Biotic Succession. Biotic succession is brought about by forces inherent within the community and in the activities of the plants and animals themselves. The most important of these activities are the organismal reactions and coactions that produce modifications in the habitat and interrelations between species. Important reactions involve filling in of ponds with plant and animal remains, the addition of organic nutrients to sterile soil, and the reduction in light intensity by increasing density of plant growth (Horn, 1971). With progressive improvement of the soil and changing light and moisture conditions, a series of new dominants come into the area. When invasion of new species occurs, intense competition develops; if the invaders are successful, the old species disappear as a new community replaces the old one.

Succession is considerably influenced by the kinds of propagules available in the vicinity. Seeds, spores and the like are dispersed more or less readily, depending on the form. Some kinds of animals roam more widely or sprend more readily into new areas than do other kinds. The composition of the community that develops and the rapidity with which it becomes established depend, in large part, on the rates at which pifferent species invade.

Bioseres may be broadly grouped as priseres and subseres, depending on whether they develop on primary or secondary bare areas. A primary bare area is a sterile habitat, such as rock, stand, clay or water. A secondary bare area is a denudation resulting from temporary flooding, fire, logging, cultivation, overgrazing, or other phenomenon that does not produce an extreme disturbance of the soil or substratum. Since in the latter the habitat has already supported community life, and since the soilsubstratum is already in an advanced stage of development, the resulting subsere progresses rapidly and the early pioneer stagess of the prisere are not usually required. A subsere will develop following the destruction of any stage in a prisere, and the species composition of the stages in the subsere will be influenced by particular priseral community that was destroyed.

The early stages or communities that make up the prisere drpend largely on the type of bare area on which the prisere originates. As succession proceeds, however, later stages in the various seres in any area having a relatively uniform and humid climate come to be more and more alike. The successional development from widely diversified communities in initially different habitats to closely similar or identical climax communities in habitats that have also become much alike is called convergence.

Bioseres may occur on a small scale in microhabitats as well as in major ones. When hay infusions, prepared in the laboratory, are seeded with representative protozoans, the order of appearance of maximum or peak populations in the various species is bacteria and monads, colpoda hypotrichates, Paramecium, Vorticella and Amoeba. Disappearance for species is in the same order, except that Amoeba precede Paramecium and Vorticella. Algae may come in at the final stage, so that a more or less balanced community is established. The succession of species results from the higher reproductive rate of earlier species and from the fact that the excreta of at least some forms, especially the hypotrichates and paramecium, are toxic to themselves.

Recognition of Communities

The organismic concept considers the community to be a superorganism, a complex organism, or a social organism. As such, it is highest stage in the organization of living matter : namely, cell, tissue, organ, organ system, organism, species population, community. There is emergent evolution, so to speak, at each higher stage in this hierarchy; the whole is more than merely the sum of its parts. The species population has inherent characteristics of density, rate of natality rate of mortality, and age distribution, while the total community has such unique functions as dominance, cooperation, trophic balance competition, and succession which are beyond the characteristics of the individual organisms of which the community is composed. The community behaves as a unit in its competitional and successional relations with other communities, in its local and geographic distribution, in its seasonal activities and response to climate, and in its evolution. Although the community varies in its taxonomic composition and structure in different environmental situations, this variation is proportionally no greater than occurs in different cells of the same type or between individuals belonging to the same species.

The individualistic concept places emphasis on the species, rather than the community, as the essential unit for analysis of interrelations, activities, distribution, and evolution. Each species responds independently to the integrated influence of the various factors of the physical environment and biotic coactions. The environment may be conceived as a pattern of gradients with the intensity of the various factors changing gradually in space from one extreme to the other.

Physiognomy. The gross structure of a community or its physiognomy is an important basis for its recognition. In terrestrial communities, physiognomy is determined by the life-forms of the dominant plant species and their specing. The life-forms that prevail in a given area depend on the climate and tne substrate or other special features of the habitat and give character to the landscape. The distribution of animal communities is closely correlated with the structure of the vegetation-types need, to be recognized and defined :

Deserty. Hot, arid habitats with scattered scrubby or thorny vegetation or, in extreme cases, none.

Thron Scrub. Tropical dry thorny deciduous tall shrubs, succulents, ephemerals.

Steppe, Plains. Semi-arid grassland covered with short grasses.

Prairie. Semi-humid grassland covered with mid and tall grasses.

Temperate Shrubland. Semi-arid areas covered with bushes and shrubs, either deciduous or broad-leaved evergreen (chaparral).

Savanna. Grassland with scattered trees or groves of trees or shrubs covering 0 to 25 per cent of the ground (Dansereau 1960).

Woodland. Open stand of small deciduous or evergreen trees covering 25 to 60 per cent of the ground and with undergrowth of grassland or desert vegetation.

Forest-Edge. Mixture of trees, shrubs and open country, ordinarily occuring as a narrow belt on the margin of forests.

Forest. Closed stand of trees forming a continuous canopy over at least 60 per cent of the area.

Deciduous Forest. Broad leaves fall during cold or dry seasons.

Broad-Leaved Evergreen Forest. No regular season of leaf fall, leaves often sclerophyllous, warm climates.

Rain Forest. Tall luxuriant forests often with several strata of trees, foliage retained throughout the year, climate continuously warm and wet.

Needle-Leaved Evergreen or Coniferous Forest. Forests of pines, spruces, firs, larches, hemlocks and the like.

Forest-Tundra. Stunted open growth of coniferous forest in cold climates

Tundra. Extensive flat or gently rolling treeless areas occurring in cold climate.

Alpine Tundra. Treeless arcas at higher elevations of mountains.

Bog. Wet areas in cold climates containing sphagnum, health, plants, coniferous trees.

Swamp. Wet areas in warm climates covered with deciduous trees or shrubs.

Marsh. Wet areas containing sedges, rushes, cat-tails and the like.

Preservation of Environme tal Quality. Up to the advant of civilized man, the biosphere of the world consisted of a complex pattern of biotic communities resposive to a vari ty of environments. The harmony prevailing between the biotic and physical realms was the result of hundreds of millions of years of evolution and adjustment of one organism to another and of each to the habitat in which it lived. The relationship was a salient one. A sporadic outbreak of a species as the result of some temporary advantage, or a catastrophe to a population as the result of a change in the physical environment, soon became tempered and a balance restored. Primitive man occupied a niche and assumed a role comparable to other animals, and the ecosystem maintained itself in a vigorous healthy condition.

Modern man, however, has initiated a number of abrupt changes during the last few centuries beyond the capacity of the ecosystem to adjust. He has become the dominant organism in the biosphere, forcing changes to fit his needs and desires, and creating ecosystems different from any that have before existed. Emergence of new communities is a characteristic of the geosere, but the replacement of the biosphere by an "anthroposphere" or "noosphere" (from the Greek word, noos, mind) is unique in the speed and extreme changes produced. This acquisition of doominancce is the result of the extraordinary evolution in man of intelligence, ingenuity, and use of tools. He has greatly modified or replaced the natural processes the control stability and balance within the ecosystem with new processes and manufactured products which the environment cannot absorb. The result has been a general deterioration of the environment, a population, explosion of the human species, and a disharmony in the world ecosystem and will, unless resolved, lead quickly to catastrophe. Whether man will survive as the dominant organism, be reduced again to the role he occupied as a primitive, or become extinct may well be decided within the next few decades. There is conflict here between the demands of man for expanding population, productivity, power and pleasures and the functioning of natural laws for maintaining stability and equilibrium between the living and physical environments.

Immediate problem of concern in the preservation of environmental quality are the elimination of air, water and soil pollution the disposal of solid wastes; and climate modification on a planetary scale. Perhaps noise pollution should also be added. Pollution is change in environment, caused by man's discharge of matter or energy into it, which alters the normal function of an ecosystem. Fundamental to the attainment of stabilized and balanced worldwide ecosystem are the control and regulation of the human population at a suitable level and preserving the habitat and the basic community of plant and animal life upon which human life depends. Considerable technical knowledge already exists for solving these problems. The immediate need is concerted attention for implementation of the necessary remedial measures. Satisfactory resolution of the problems will involve not only science and technology but also law, sociology, politics and economics. Emphasis on existing knowledge, however, should not obscure the fact that continuing basic research is required to elevate man's understanding of the environmental system. Scientists are just beginning to study ecosystems on the multidisciplinary basis that is clearly required.

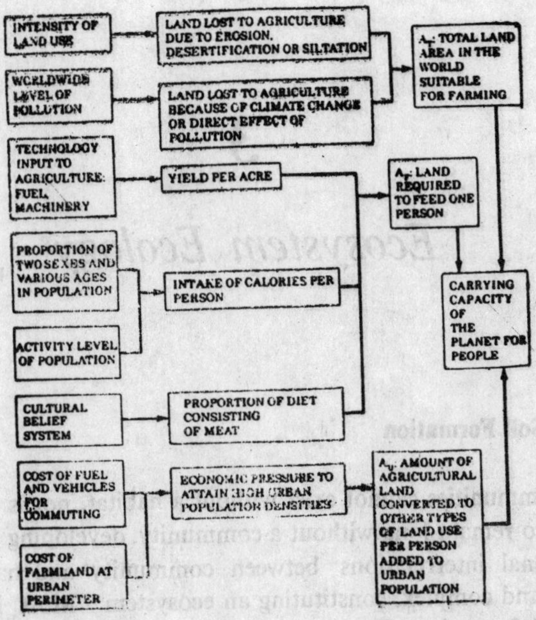
INTENSITY OF LAND USE
LAND LOST TO AGRICULTURE DUE TO EROSION, DESERTIFICATION OR SILTATION
WORLDWIDE LEVEL OF POLLUTION
LAND LOST TO AGRICULTURE BECAUSE OF CLIMATE CHANGE OR DIRECT EFFECT OF POLLUTION
TECHNOLOGY INPUT TO AGRICULTURE: FUEL, MACHINERY
YIELD PER ACRE
PROPORTION OF TWO SEXES AND VARIOUS AGES IN POPULATION
INTAKE OF CALORIES PER PERSON
ACTIVITY LEVEL OF POPULATION
CULTURAL BELIEF SYSTEM
PROPORTION OF DIET CONSISTING OF MEAT
COST OF FUEL AND VEHICLES FOR COMMUTING
ECONOMIC PRESSURE TO ATTAIN HIGH URBAN POPULATION DENSITIES
COST OF FARMLAND AT URBAN PERIMETER
A_{T}: TOTAL LAND AREA IN THE WORLD SUITABLE FOR FARMING
A_{F}: LAND REQUIRED TO FEED ONE PERSON
CARRYING CAPACITY OF THE PLANET FOR PEOPLE
A_{U}: AMOUNT OF AGRICULTURAL LAND CONVERTED TO OTHER TYPES OF LAND USE PER PERSON ADDED TO URBAN POPULATION

3

Ecosystem Ecology

(A) Soil Formation

Communities cannot exist without a habitat, nor is a habitat likely to remain long without a community developing in it. The functional interrelations between community and habitat are many and complex, constituting an ecosystem. Most important are soil formation nutrient cycling, and energy flow. Human interference in these processes often causes pollution, and exploitation, often bring exhaustion of natural resources.

We have already considered many of the reactions of plants upon the habitat, such as reduction of light and wind intensities, mitigation of temperature extremes, interception of rainfall and increase in relative humidity. Plants also exert important effects on the formation, structure and characteristics of the soil or substratum produced by accumulation of dead plant remains; they further the weathering of rock through acid excretion and the mechanical action of roots; they offer obstruction to wind and water borne materials; they help stabilize moving sand and talus slopes and help prevent erosion generally; they variously increase or decrease the water content of the soil; they foster decomposition of raw humus into usable untrients; and so forth. Water plants form marl. It is by these reactions that plants exert dominance interrestrial communities and establish the physical conditions of the habitat, which must be acceptable to all minor plants and animals that dwell there. Succession of plant stages eventually brings the interactions between habitat and community into equilibrium upon the establishment of the climax. In this chapter we will be primarily concerned with soils as dynamic components of ecosystems, including the cycling nutrients between the soil and the biota.

Soil Formation

Texture porosity, consistency arrangements of particles, chemical nature, and organic content of the soils are determined by three factors; parental rock material, the biota, and the climate. Differences in topography modify the relative effects of these three factors, and time is required before their full effects are realized.

Parental Rock

The basic rock from which the mineral portion of a soil is derived determines, to a large extent, not only its chemical composition but also its structure. For instance, soils derived from limestone are highly calcareous and more alkaline than soils derived from limestone are highly calcareous and more alkaline than soils derived from feldspar, sandy soils from quartite. Clay forms a finely textured compact, water-retaining soil. Sand is coarse-textured and porous. Loan is a mixture of sand and clay and makes the best soil. The presence of iron oxides and silicates produces the red and yellow colors of some soils. Humus produces black soils. Soils from swamps or bogs and very rich in organic material is called muck.

Residual soils are formed in situ from underlying bedrock. Soils may however, be formed in one locality and moved considerable distances. Soils transported and deposited by wind are called losses, by water alluvium; by glaciers, till.

Biota

Plants and animals have a highly important role in the formation of soil, both of them affect its structure and they aid in the production of humus. Plants contribute to the mechanical and chemical weathering of rock. Plant roots, especially those of trees, can split large rocks. Lichens, mosses, and even bacteria and fungi excrete acids in the course of metabolism, which dissolve the substances that cement rock granules together. When plant roots die, fungi convert them to dry, soft spongy materials (punk), used as food by saprophytic micro-arthropods. Usually the bark of the root remains in tact the longest. Hollow tubes are thus formed that permit water and air to penetrate considerable depths into the soil. These channels gradually become filled with silt and animal excreta.

The addition of plant and animal organic matter to heavy compact soils or clay tends to open the soil, making it more porous. Addition of organic matter to sandy soils binds the particles closer and together, making the soil less porous.

Earthworms may be divided into deep and shallow-working species. Deep working species dig narrow tube-like channels which may reach 2-3 metre down through overlying soil to parent rock. Earthworms ingest soil while borrowing, digest and absorb organic matter from it, and digest the residue in semi-liquid form which is used to cement the walls of the burrow or else is deposited at the surface as castings. Earthworms prefer easily digested succulent vegetation and dung for the purpose, but in the autumn may pull the freshly fallen leaves down into their burrows to use as food or nest linings. Ejected petioles may form midden piles around burrows. Earthworms are not, however, important soil builders in disturbed grassland; they may be absent altogether in and regions, the dry weight of casts brought to the surface annually by earthworms varied from 475 g/m^2 (2.1 tons/acre) in a moderately hot dry climate, to 24,000 g/m^2 (10^7 tons/acre). Earthworm casts compared with the surrounding soil show higher total nitrogen, oganic carbon, exchangeable calcium, exchangeable magnesium, available phosphorous, exchangeable potassium, organic matter, base, capability, pH and moisture equivalent. Only certain species make these surface castings; other species void the ingested soil into subterranean spaces.

The ant Lasius niger neoniger spends most of its time in its underground burrows and deposits excavated soil upon the ground surface around burrow entrances.

Humus in soil, organic matter that is partly or entirely decomposed is called humus. The amount of humus varies from less than 1 per cent to as much as 20 per cent of the soil; peat soil may be largely organic material, but much of it resists decomposition, and hence is not true humus. Decomposition breaks down compounds into similar ones that are washed back into the soil, thus becoming available again as nutrients.

The annual dry weight of leaves that fall to the ground in deciduous and coniferous forests varies from site to site and with the density of the trees, but is commonly in the range of 50 to 400g/ms². In mature climax forests the rate of decomposition of the liter and re-absorption by plants of the nutrients thus yielded keeps pace with the annual accumulation so that an equilibrium is established. In several stages, decomposition and utilization do not keep up with the annual accumulation, so that the ogranic content of the forest floor increases with time. Litter production is lower in artic than in tropical regions.

Metric Tons/Hectare-year

Region	*Leaves*	*Other*	*Total*
Arctic alpine	0.7	0.4	1.0
Cool temperature	2.5	0.9	3.5
Warm temperature	3.6	1.9	5.5
Equatorial	6.8	3.5	10.9

However, in the warm tropical regions the amount of humus that accumulates on the forest floor is low because of the high rate of decomposition, runoff and leaching.

The thick organic layer on the ground moderates extremes in the daily and seasonal rhythms of soil temperature, retards freezing of the ground in the autumn and thawing in the spring, and retains soil moisture. Because of humus formation (involving oxidation) and the respiration of plant parts and animals undergrounds, soil air contains little oxygen but much carbon dioxide, and it possesses a higher moisture content than does the general atmosphere above ground. This is especially marked in warm summer months when these processes go on more rapidly. The decay of organic matter usually makes the top soil somewhat acid (most commonly pH 5 to 7), but in the mineral subsoil, the acids are often neutralized by the basic salts commonly present.

The mineral content of leaf fall varies according to the species of tree.

Element	*Ha dwood forests*	*Coniferous forests*
Cal.ium	7.3	3.0
Ni rogen	1 8	2.6
Potassium	1.5	0.7
Magnesium	1.0	0 5
Phosphorus	0.4	0.2

Silicon, cooper, manganese, carbon and zinc are also present in the leaves of hardwood trees. Carbon is relatively more abundant and nitrogen less abundant in coniferous than in deciduous leaves, but commonly the carbon nitrogen is 55 : 20.

Both plants and animals are important agents affecting the decomposition of organic matter and the formation of humus. An animal digests and mobilizes plants foods, much of which is returned to the soil, in parts as the excreta of the living animal, in part as the body of the dead animal, in part as gas. Fully formed human is in fact, derived mostly of fecal material. The larger herbivorous and carnivorous animals pass urine and feces containing simple nitrogenous compounds and compounds of phosphorous, potassium, and traces of calcium, magnesium, sulphur and other elements. Humus is but one point in a continuous cycles of decomposition of plant and animal organic matter, absorptian of decomposition products by plants ingestion and metabolization of plant matter by animals, decomposition of plant and animal organic matter.

The most important soil organisms concerned in the decomposition of the little are the bacteria, both aerobic and anaerobic forms. T ey are commonly divided into two types. Hetrotrophic bacteria obtain their energy from the oxidation of the carbohydrates and fatty substances as described above. They use this energy for the synthesis of cell substances and the production of enzymes that break down the complex compounds in the litter into simpler compounds, including proteins into ammonia compounds. They then use part of the ammonia compounds in synthesizing the amino acids they need in building their own proteins. Autotrophic bacteria, in turn, are of two types : chemosynthetic species, which obtain their energy from the oxidation of inorganic compounds (hydrogen, sulphur, hydrogen sulphide, iron and ammonia) and photosynthetic species which include purple and green sulphur bacteria, possess a form of chlorophyll, and utilize the energy of sunlight. Chemosynthetic bacteria convert ammonia compounds into nitrites and nitrates, part of which they use in their own anabolism, the rest becoming available for plant to absorb. Photosynthetic bacteria use the ammonia compunds in their own anabolism and do not render them directly available to plants. Chemosynthetic bacteria are more abundant than photosynthetic bacteria in soil; photosynthetic bacteria are more abuntant.

The decomposition products available to plants are reabsorbed by the roots and built up into plant tissue. Plants are eaten by animals and as plants and animals die the minerals are returned to the soil. All essentiel minerals cycle repeatedly through the ecosystem, the cycle of each mineral differing in various ways from the cycle of every other mineral.

Major Soil Groups

The interrelations between the basic mineral content of the parent substrate, biotic reactions, and climate can be seen through an analysis of the development of the great soil groups of the world. We will give only a simple classification of them. A detailed classification would include many subdivisions and intermediate categories (Lutz and Chandler 1946, Simnoson 1957).

Podzolic soils are formed in humid temperate climate, under forest vegetation. The A_2 horizon is moderately well developed, for there is sustained leaching. Soils are more or less acid and only moderately fertile. Podzols develop under coniferous forest and have a mor type of humus. Graybrown and brown podzolic soils are found under hardwood forests and have a mull humus.

Latosolic soils develop in humid tropical or semitropical forested regions. Humus is quickly oxidized by action of micro organisms and hence does not accumulate. The soil fauna is depauperate. Chemical weathering of the parental materials is intense. Water drainage through the porous soil is rapid, so eaching is extensive. In early stage of its formation, the soil is neutral or slightly alkaline, but as leaching continues, it becomes acidic. The soil has a thin organic layer (A_0 and A_1 horizons) on a reddish, leached soil (A_2 horizon) that extends to great depths below the surface.

Chernozemic soils occurs in humid to semiarid temperate climates under grass vegetation. The grasses on dying return considerable organic matter to the soil. The A_1 horizon is consequently dark in color and of great thickness. The soil contains more bases and hence is less acid than in the two types above. The B harizon in humid regions is indistinct, but where there is less rainfall, calcium salts may accumulate to form a hardpan. Prairie soils in temperate climates are among the most fertile soils of the world, but fertility decreases in the tropical and desert climates.

Desertic soil are characteristic of arid climates and contain very little organic matter. A profile is poorly developed. The surface soil is brownish gray, and grades quikly into the calcium carbonate horizon which usually forms a hardpan just below the surface. Wind erosion removes the finer particles leaving the coarser materials to form a hard pavement. The soils are but slightly weathered and leached; lacking nitrogen, they are infertile.

Mountains and *Mountain Valley solis* vary shallow layer or eroding rocks to deep organic soils of valleys and swampy areas.

Tundra soils occur in cold northern areas where the substratum remains continuously frozen and the vegetation of lichens, mosses, herbs and shrubs makes a peaty surface layer. The region is poorly drained and characterized by many scattered shallow ponds.

Alluvial soils may be important locally. These soils are mostly without a developed profile and are the result of deposition by streams. The are usually very fertile and support luxuriant vegetation.

Saline soils are found in dry climates where rapid evaporation of water results in surface deposition and accumulation of salts leached from surrounding upland areas.

This classification shows clearly the correlation that exist among soil groups, climates, and vegetation. The compositiont and particularly the abundance of animals is also affected. The average biomass of soil animals under different categories of vegetation-soil-climate system is as follows (Ghilarov 1967) :

Moss tudra : 3 g/m^2

Coniferous forest : 20 g/m^2

Broad leaved deciduous forest : 100g/m^2

Short grass plains : 25 g/m^2

Semi-desert : 1 g/m^2

Soil Erosion

As rain water drains over the surface of the ground, it picks up and transports particles of soil more or less proportional to the rate of flow. This is a continuous process that has gone on throughout geological time, resulting in base-leveling of mountains, the filling in of valleys, and the formation of river deltas. Normally this is a slow process, with changes becoming apparent only in thousand years. With the ground covered with vegetation, the flow of water is retarded and more of it percolates into the ground or is evaporated. In fertile areas, new soil is formed faster than it is eroded or at least there is stabilization.

With removal of the native vegetation and cultivation of the soil in agiculture, surface particles are more easily moved and surface water drainage is more rapid. As a result, topsoil is being lost at an alarming rate. Much land has been so badly damaged that here is little likelihood that it will ever again be productive crops. Conservation of the fertile top soil is one of the major problems facing modern agriculture. Putting in underground drainage tiles in level farmlands helps to draw rain water into the soil and to decrease surface run off and erosion. Soil conversation procedures should be strictly followed.

(B) Food and Feeding Relationship

Food-getting necessarily involves interrelations between organisms and between species; these interrelations are among the most important coactions in any community. Animals are adopted variously to capture and utilize certain types of food and to avoid being captured by other animals. One must understand these adaptations and interrelations to appreciate properly the role that food-getting plays in the dynamics of the community.

Feeding Behaviour

Free-Living animals are commonly classified on the basis of normal feeding behaviour, thus :

Herbivores : Feed principally on living plants.

Carnivores : Feed principally on animals that they kill.

Omnivores : Feed nearly equally on plants and animals.

Saprovores : Feed on dead plants and animals, and excreta.

The various categories are capable of further subdivision. Thus herbivores includes (a) those that feed chiefly on the foliage as the large cursorial grazers, such as bioson, antelope, muskox, caribou, sheep; small surface, living grazers such as rabbits, mice, grasshoppers; subterranean-living grazers, such as wood-chucks, prairie dogs, kangaroo rats, ground squirrels; browsers, which feed on foliage buds and twigs of trees and shrubs rather than strictly on grass or ground herbs, such as wapiti, deer, moose, grouse, and defoliating types of insects such as the hemlock looper, spruce budworm and larch sawfly; (b) those that feed chiefly on the products or non photosynthetic parts of the plants as seed nut and fruit-eaters, such as squirrels chipmunks gallinaceous birds, sparrows, plant-juice suckers, such as aphids leaf-hoppers, mosquitoes, chinch bugs, and cambium feeders, such as bark bettles gall flies, cympids : and (c) those that feed primarily on fungi and bacteria fungivores. Some forms that appear to be saprovores are actually fungivores, feeding on the fungi or bacteria that they obtain from decaying organic material.

Carnivores are also predators. Carnivores restricting their food chiefly to insects are called insectivores : those limiting themselves largely to fish are called piscivores; and so on. Parasitoids eventually consume their hosts, and hence are a special type of carnivore. Some plants are carnivourous. The picher plant. Venus's-fly-trap, and sundew, which grow in bogs or wet places, and bladderwort, which occurs in ponds, depend for their nitrogen supply largely on animals that they capture and consume. Some soil fungi trap and consume nematodes. Perhaps the bacteria fungi, and viruses that cause disease in animals also belong to this classification.

Many species eat both plant and animal matter, on occasion, or at particular season, but animals are considered to be truly omnivorous only if they feed on plants and animals in nearly equal amounts or indiscriminately. Omnivores occasionally also consume dead organic matter. Some aquatic organisms are filter-feeders and may consume everything within a particular size and range that passes through their feeding apparatus. However, filter-feeders may demonstrate selectivity by feeding in neighbourhoods where certain species predominate. Some copepods select particles of a particular size, rejecting larger ones, regulating the distance between the maxillae and the filter mechanism.

Feeding Adoption

Among kinds of mammals, teeth show considerable adaptive radiation, correlated with the type of food consumed. The molar and premolar teeth of insectivorous species, such as shrews and bats, are low and have sharp pointed cusps for crushing weakbodied prey. The piscivorous toothed whales have largely lost all differentiation in their teeth which are simple conical, grasping structures. The teeth of the carnivorous dogs and cats are high-crowned and tubercular, well fitted for shearing flesh. Herbivorous ungulates and rodents have teeth that are flat-crowned, suited to grinding harsh grasses and other vegetation. Their jaws are capable of considerable lateral motion. Omnivores may have both grinding and pointed teeth. Saprovores have rather blunt teeth. The ant-eating sloths and their relatives have no teeth, and the mouth is almost tubular in shape. The tongue has become long and prehensile for lapping up tiny ants.

The bills of birds display great variety in shape and size, adaptations to feeding in numerous quite specialized niches in the environment. The tongues of birds are variously modified to serve as long probes or spears (woodpeckers and nuthathches), as a strainer (ducks) as long capillary tube for obtaining nectar from flowers (humming-birds,) as a rasp (hawks and owls), as a finger for manipulating the food in the mouth (parrots and sparrows), and as a tactile organ (sandpipers and herons).

The mouth parts of insects are adapted primarily either for biting and chewing or piercing and sucking. Among marine invertebrates, adaptations for feeding on detritus include pseudopodia (Foraminifera) ciliated epithelium that maintains a flow of water through the animal (sponges, calms); prehensile, often ciliated arms (various polychaetes; holothuorideans) and soft eversible gullets (various polychaetes, sipunculids). Those that are herbivorous or carnivorous, as well as detritus feeders have prehensile tentacles armed with nematocysts (mollusks), eversible stomachs (starfish) and masticatory structures in the mouth or stomach, (crustacea, diptera larvae)

In addition to mouth-par adaptations, there are many modifications in other parts of the digestive tract for handling particular types of food. These adaptations occur throughout the animal kingdom, but are especially evident in birds. A crop is present in some species but not in others. The walls of the stomach are more muscular in seed-eating birds than in flesh-eaters. Owls and some other species form and regurgitate from the stomach pellets of indigestible matter. Gallinaceous birds are able to retain or shift the supply of gravel or grit in the stomach as an aid to grinding seeds. The length of the intestine varies with the type of food consumed; the caeca are longer in browsing and seed eating gallinaceous birds for digesting cellulose.

Associated with these anatomical and histological adaptations are adjustments in function and behaviour. Obviously, if an animal has morphological adaptations for ingesting and digesting flesh, it must behave as a carnivore and not as an herbivore. The possession of these adaptations and adjustments means that animals are generally restricted to the particular type of food that they can use most efficiently. The kinds of food eaten by animals. is of fundamental ecological and economic importance.

Choice of Food

The kinds of food eaten by animals depends on their genetic heritage and parental training or conditioning while young. Involved in the evolution of the food habits of a species and the "profitability" of consuming one prey species rather than another are the animal's physical adaptations for ingesting and digesting particular types of food, the nutritional values of the food. Its palatability the size of it, its availability, or abundance, a:d its ease of procurement, which depends in large on the various protective devices that it possesses.

Nutritional Values

Animals generally require proteins, carbohydrates, fats, vitamins, minerals and water. Proteins are used as the basic substance in the composition of protoplasm; carbohydrates and fats are oxidized to furnish energy for the body; vitamins serve as catalyst for specific metabolic processes, minerals are needed to regulate osmotic pressure and as constituent elements of various body organs; and water is used as a general solvent, lubricant, and circulatory medium. Species differ, however in their needs for particular substances. The beetles Tribolium, Lasioderma, and Ptinus, for instance, grow slowly but nonetheless satisfactorily on diets lacking carbohydrates. Hence, they may be disturbed more widely than are species which require carbohydrate in their diet.

Foods differ in composition. Foods staple to an organism's diet are those easily digested and of high caloric and protein content. They are adequate for sustenance of the weight and vigor of the animal, but usually need to be supplemented with vitamins and minerals. The bob-white and ringnecked peasant, for example eat certain cultivated grains and weed seeds, such as corn, sorghum, barley, wheat, rye, soya beans pigeon grass, and lesser ragweed, as staple foods, at least on a mixed diet.

Non-staple or emergency foods are not in themselves sustentative, and animals limited to them gradually lose weight and die. Such foods are, howerver, often abundant and easily procured in emergencies. When staple foods are covered with snow or ice and furnish sufficient energy to tide the animal over the critical period. In emergencies, the bobwhite and ring necked peasant eat black locust beans, fruits of bittersweet and sumac, rose hips, dried wild grapes, and sweet clover seeds.

Vitamins are necessary for the maintenance of good health in world animals, just as in domestic animals or man. The symptoms of vitamins deficiency, induced experimentally, are similar. Evidence has been difficult to secure, however that animals suffer from vitamin deficiencies in their natural environments.

Animals obtain most of their required minerals from their food and water. Additional salt must sometimes be given to caged animals to prevent cannibalism. The gnawing of cast off deer antlers by rodents is apparently for additional salts. The use of certain soil deposits and springs as natural "licking sites" by deer and other ruminants is apparently for sodium salts lacking in their general diet. Some birds, such as the evening grosbeak are also attracted to sources of salt supply. Grit in the form of gravel, sand, or small stones in the stomach of gallinaceous and other birds may not only aid in the grinding of seeds and hard vegetables matter but also be a source of minerals.

Digestibility

Different species of animals very considerably in efficiency of digestion and utilization of particular food substances. Thus, clothes-moth larvae can digest cloth and bird lice can digest feathers, because among other things they have an exceptionally high hydrogen ion concentration in their intestines.

Digestive enzymes occur generally throughout the animal kingdom although less is known about them in the protozoa. In the lower phyla the enzymes are generalized in respect to the kinds of foods on which they act; in the higher phyla, they become highly specialized. A specific enzymes, however, does not differ greatly from one animal group to another. Carnivores have strong, proteases and weak carbohydrates, correlated with their meat diet. Herbivores on the other hand, have weak protein, but active carbohydrate, enzymes. Herbivorous mammals and birds posses a bacterial flora in their digestive tract that makes possible digestion of cellulose. Omnivores have a full complement of enzyes and can utilize a wide variety of foods.

Practically all food contains some indigestible matter, which is ordinarily passed through the digestive tract and eliminated in the feces. If the indigestible material is excessive, or if it contains toxic substances, regurgitation or vomiting may result.

Size of Food Item

The size of the food in relation to the animal is not of major importance to many herbivores or saprovores as they normally feed in or on the organism or substance. With carnivores, the size of the prey must be within their power of conquest. Ordinarily the size of prey is less than that of the carnivore that feeds on it, but a high degree of ferocity and audacity, or pack hunting in the manner of wolves, often enables the carnivore to take prey larger than itself. On the other hand, a predator cannot profitably prey on species so small that the energy derived from its consumption does not equal the energy expended in its capture. Evolution of prey species subject to heavy predation tends to proceed either to very large size or very small size in relation to the size of their most important predators. Some very large aquatic predators, however, have become adopted to feed with a minimum effort on very small organisms occuring in dense concentration through the evolution of a filtering appartus in their mouth parts. A good example is the feeding of blue whales on plankton.

Availability

In order to determine if a species is fed upon in population to its abundance, it is necessary to find out what animals have been eating of that which is available in a habitat. The relation between the two may be shown graphically and expressed as forage ratio per cent of species in animals' food divided by per cent of species in habitat. A value of unity indicates that the food item is taken in proportion to its abundance, a value greater than unity indicates that it is taken more frequently, values of less than unity indicate that the item is either inaccessible, of the wrong size, too difficult to obtain, or is actually avoided. It is apparent that while there is a relationship between the relative abundance of various species and the degree to which they were taken as food, there are also several discrepancies.

The vulnerability to a predator of a prey species is directly proportional to its relative abundance among the other species available. Voles are preferred by the red fox over white footed mice, but both mice and voles are preferred over moles, shrew, and snakes. Predator preferences among them, however, occur only when all are abundant. When prey species are reduced in numbers and difficult to get, little or no predator preference occurs.

There is some evidence that predators-particularly insectivorous birds - when searching for prey concentrate on one or a few species at a time. By a kind of learning, they acquire what may be called specific searching images for these species and thereby mostly disregard other species. When a new species becomes numerous in an area they feed on it at first only as the result of chance encounters. To obtain a preference for the new species, they must become conditioned to it gradually and must learn the proper cues of where and now to search it out

Predator species that remain in one habitat throughout the year must adjust their feeding to the kinds of food available in each season. Species that are unable to do this are compelled to migrate, hibernate, or make other adjustments to survive the unfavourable seasons. Thus birds that are strictly insectivorous may occur in a given area only during the warm part of the year, leaving it before insects disappear. Omnivorous species commonly changes from an insect diet in summer to diet of seeds and fruits in the autumn and are often non-migratory.

Protective Devices

Most, if not all, plants are equipped with defence mechanisms against the grazing and browing of animals. The most widespread defences are chemical – various kinds of better of toxic materials in plant tissues that make unpalatable to many animals. Animals evolve tolerances and neutralizing devices, however, so that most plants are fed on; their defence is relative, not complete. New leaves often contain less of the defensive chemicals, and are thus more vulnerable. Trees and shrubs may be deformed or killed be excessive browsing that destroys terminal twigs and buds. Grasses are not so affected, for they grow from the base of the leaves; they may even be benefited by moderate cropping Grasses also are protected by grains of silica embedded in their tissues; only by evolution of high-crowned teeth adopted to abrasion by the silica have some ungulates become able to graze on grass. Some species of trees, shrubs, and cacti are protected from browsing by prickles or thorns. The protection is important to the plants for survival in deserts and in grazing subseres of humid regions. Some plants are particularly noxious or toxic and animals quickly learn to avoid them. Seeds may develop alkaloids or other disagreeable substances that give them protection.

Various kinds of arthropods have delcusive secretions in the form of sprays that they emit when attacked. There is increasing evidence that attractment and repellant secretions are important in both predator prey and social relation of animals.

When coloration renders it inconspicuous in its normal environment, an animal is said to have concealing or cryptic coloration. When the coloration or markings reproduce the general tone or characteristics of background, it is called protective resemblance; disruptive, when the markings break up the outlines of an animal and replace it by some irregular configuration so that the animal is not recognized as prey.

These are many examples of mimicry among insects. Batesian mimicry is the resemblance of a palatable species in external features to an unpalatable one that in turn possesses warning coloration and is the more abundant of the two species. The palatable species derives benefit from the relation, since predators, specially birds avoid them as well as the unpalatable ones. The effectiveness of the mimicry decreases as the number of mimics equals or exceeds the number of models. There is considerable controversy about mimicry, however, and even the classical example of the butterflies is disputed. In Mullerina mimicry, both model and mimic are unpalatable. Pooling of numbers between the two species give more chances for inexperienced birds to lean to avoid them and reduces the losses per species during the learning process. The yellow jacket and polisitine wasps are examples and many more can be found among butterflies.

Some animals posses bright spots of colors so placed on the body as to be deflective. The attention of pursuing predator is drawn to less vulnerable parts of the body; for instance, eyespots on the fins or tail of the fish. Eyespot on the wings of some butterflies and moths are concealed at rest, but when flashed out, spread wings may frighten away an attacking bird or other predator.

Measurement of Productivity

Productivity may be measured during any reasonable period of time. Because of essential metabolic differences between day and night, however the 24-hour day is the smallest practicable unit. Similarly, because of seasonal changes in the environment and in community populations, the measurement of annual production is probably most useful. Since primary production is basic and concerns the capture of energy by plants, it will be considered first.

Primary Production

Various methods are employed for measuring primary productivity, each procedure having certain advantages and disadvantages (Ryther 1956, Mann 1969). Further work in evaluating and improving these methods or developing ones is desirable.

A common procedure for analyzing aquatic habitats, dating back to Gaarder and Gran (1927), is to suspend during daylight hours equal samples of green phyto-plankton, ordinary in separably mixed with bacteria and zooplankton, in both transparent and blackened bottles at the same depth at which obtained. Photo-synthesis of course does not occur in the blackened bottle, and there is loss of oxygen, resulting from respiration, R, and decomposition, E + D + W. In the transparent bottle, photosynthsesis occurs in addition to respiration and decompostion, bringing a production of carbohydrates. There will either be an increase in oxygen concentration, or the loss of oxygen will not be so great as in the blackened bottle. The difference in the final oxygen content of the two bottles will be a measure of gross production : I_n.

If the oxygen content of the water is measured at the begining of the experiment, the loss of oxygen in the blackened bottle substracted from the difference in oxygen content of the two bottles at the end of the experiment will represent the net productivity. This net productivity may also be determined from the difference in the oxygen content of the transparent bottle between the begining and the end. The obtain net production for an entire dailycycle, the consumption of oxygen for respiration and decomposition over 24 hours must be subtracted from the gross photosynthetic output during daylight hours.

One may use the amount of carbon dixide absorbed during a period of times as a measure of photosynthesis if correction is made for the carbon dioxide given off in respiration and decomposition. Changes in the amount of CO_2 in the water may be calculated from the differences in pH of the hydrogen-ion concentration. As an illustration of the use of radioisotopes, net production during the daylight thours may be measured by introducing a known amount of $11CO_2$ into a volume of water where the amount of carbon dioxide already present is known. The amount of 11C absorbed by the phytoplankton can be accurately determined by use of counters applied to phytoplankton collected and dried at the end of the period. Then the proportion of the radioactive carbon absorbed to the amount introduced can be applied to the total CO_2 initially present to get the total amount absorbed.

Since nitrogen and phosphorus are metabolized more rapidly by plants in the manufacture of food during the growing season than they are regenerated from decomposing material, the rate and extent of the depletion of nitrates and phosphates is freely circulating bodies of water or in the epilimnion of stratified lakes serve as an index of an amount of organic matter produced. The rate of accumulation and regeneration of these substances is the hypolimnion from the dead organisms that sinks into it may also be used to get an approximation of primary production. These measurements are not exact since they do not account for the repeated regeneration and neutrilization of the substances in the photic zone during the season or their transference to and storage in the bodies of animals.

The rate of photosynthesis varies in relation to the amount of chlorophyll present and to light intensity. The amount of chlorophyll in the standing crop phytoplankton may be determined photometrically for all depths and calculated in terms of unit area of surface (Ryther and Yentsch 1957, Flemer 1959).

Determination of the rates of photosynthesis and oxygen use in streams and ponds may be made by direct measurement of changes in the concentration of oxygen and carbon dioxide in the water as between day and night. Because of photosynthesis, there is no increase in oxygen concentration during the day time. At night the oxygen loss gives a measure of the rate of respiration and decomposition and this presumably remains the same throughout the 24-hour daily cycle. Adding the average hourly night loss to the average daily gain during the day and multipling by the hours of daylight gives the total gross production for the 24-hours day. To obtain the net production for the entire day, the hourly loss at night must be multiplied by 24 and subtracted from the total gross production. Corrections need to be made, however, for possible diffusion of oxygen from air into the water, particularly at night when oxygen concentration in the water is lowered, and diffusion of the water during the day, if supersaturation occurs. Additional correction will also be necessary for import of oxygen from groundwater and surface drainage and export of oxygen and carbon dixoide downstream by swift currents.

Secondary Production

When an animal species is represented by a low overwintering population, or an immature stage, the maximum biomass obtained in each generation this the approximate not production is for that generation. However, is does not account for continued reproduction and growth of individuals after the maximum biomass of the population is attained, nor does it account for excreta natural deaths, or the kill of predators. If the population of the species is maintained at a more or less uniform level throughout the year, the mean biomass times the number of generations gives the net production, again with the exception of the factors mentioned above. Linderman considered the phytoplankton turnover, or the production of a new generation, to occur every week form May to September and every 2 weeks through the rest of the year, the zooplankton to replace itself bi-weekly through the year, Chaoborus to have three generations per year; midge flies, two; and various acquatic bettles and bugs, one generation per year. Juday (1940) estimated that the mean standing crop of both phytoplankton and zooplankton replaced itself every 2 weeks throughout the year. To obtain gross productivity, the respiratory rates of these animals must also be measured.

Although measurement of energy flow through the total phytoplankton, zooplankton, and smaller soil organisms is often practical and sufficient, measurements are required on individual species of larger size in the higher tropic levels before the total utilization of energy by the tropic level can be determined.

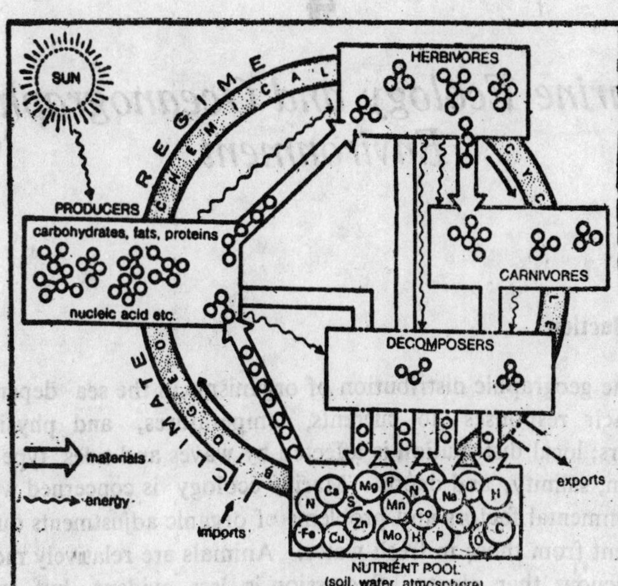

A generalised model of an ecosystem to show its structure and function. Note, the biogeochemical cycles superimposed on the various components of the ecosystem; movement of materials in cyclic manner, and that of energy in non-cyclic (unidirectional) manner.

4

Marine Ecology and Oceanographic Environment

Introduction

The geographic distribution of organisms in the sea depends on their responses to currents, temperatures, and physical barriers; local distribution is affected by waves and tides type of bottom, salinity, and depth. Marine ecology is concerned with environmental factors and problems of organic adjustments quite different from those in fresh water. Animals are relatively more conspicuous than plants. Succession is less evident, but such ecological processes represented by chemical cycles, co-operation and disoperation, food chains, productivity, population, dynamics, niche segregation, speciation, and dispersal are fully as import nts as on land.

Distinct self-contained community units are more difficult to recognize in the sea than on land because of the apparently greater interrelation of benethic species and the free movement with circulating currents of plankton and nekton. Plankton is everywhere a basic link in food chains, but the general distribution and importance of plankton species in the sea is no more remarkable than that of soil organisms in terrestrial biomes. To consider the entire ocean community as a single biome, as has been suggested by some investigators, is stretching the concept beyond its usefulness. Since we identify biomes by differences in the life-forms, and functional adjustments of the conspicuous dominant or predominant organisms, we may properly recognize biomes that occur in the open ocean, on eroding rocky shores, on muddy and sandy beaches, and composing the coral reefs and atolls. Each of these biomes may be subdivided by the taxonomic composition of the predominant organisms into secondary communities equivalent to the biociations that we have recognized on land. Much of the early literature on marine communities has been reviewed by Gislen.

Habitat

The marine biocycle is considered to have benthic (bottom) and pelagic (open water) divisions, the littoral zone of the ocean shore extends between the limits of high and low tides, the sublittoral zone covers the continental shelf to a depth of about 200 m. the approximate depth at which maximum wave action produces any effect. The average depth of the ocean is about 3800 m. but oceanic trenches (hadal zone) extend much deeper, the Marianas Trench in the Pacific Ocean to approximately 11.600 m. The neritic biochore is above the continental shelf and is commonly 16-240 km (10-150 ml) wide. The oceanic biochore is subdivided vertically with the boundary between the epi-and mesopelagic zones, depending on the extent of effective light penetration.

Tides

The level of water in the ocean rises and falls usually twice each day or at an interval of 12 hours and 26 minutes. In some parts of the world the tides are less regular or there may be but one daily. Flood-tide is the period in which the level is rising and covering more and more of the shoreline; ebb-tide is the period in which the water level is less than a meter, but the change may be much more than this on the shore, depending on its configuration, the Bay of Fundy opens broadly to the sea and tapers to a narrow head landward, and tides may be 6 to 10 or even 15 m. On the other hand, when bodies of water have only a relatively narrow connection with the sea, as does the

Gulf of Mexico with the Atlantic, the range in water level is less than 30 cm. Even lakes have seiches, but they are hardly perceptible except in the larger lakes, where it may amount to a few centimeters.

Tides are caused by the attraction of the moon and, to a lesser extent, the sun. When the sun's attraction is added to that of the moon, as occurs twice each month at the time of full moon and new moon, the fluctuations of the tides are unusually high and unusually low, these are called spring tides. When the tidal influences of sun and moon are opposed, as happens twice each month, the tides have the least amount of flow and ebb and are called neap tides.

Tides have their greatest effect on animals on the seashore, because of the associated pounding of waves and the alternative submergence in water and exposure to the air. However, the organisms appear well adjusted to this rhythmic submergence and exposure (Flattely and Watson 1922, Korringa 1947). For instance, as the stones on which the chiton occurs become exposed, the animals react positively to gravity and negatively to strong light, and move downward. They travel at maximum speed while the stone is still moist and become aggregated on the damp lower sides of the stones. When the stone again becomes immersed by the returning tide, the animals lose their geotactic orientation, and, since, illumination becomes more or less equal on all sides of the stones, they move about at random until they reach the upper surfaces again.

Substratum

The pounding action of waves on rocky shore may have tremendous force, estimated in one instance at 15,000 kg/cm^2; animals occupying exposed rocky shores in the surf belt must be strongly protected and firmly attached (Flattely and Walton 1922). The conical shaped limpets present a minimum of surface to the waves. Large depressions in the rock retain water at ebb tide to form tidal pools and thus may contain the more delicate species because of the protection they afford. The various sea weeds absorb some of the wave shock for the animals living with them. The shape, form and size of corals, sponges, and other colonial type are affected by the amount of wave action to which the animals are exposed, a shall of the Mytilus mussel may weigh 58 g where the animal is exposed to a heavy surf but only 16.5 g in more quiet waters.

Sandy beaches occur only where the force of waves is reduced by being spread over a more gentle slope. Even here, especially during stroms the sand makes a very unstable substratum and not many animals except mollusks and some of the echinoderms can keep from being smothered or buried. Mud buttoms occur only in relatively quiet waters. Burrows made in mud hold their form better than in sand, so large populations of animals can occur in mud.

Organic deposits are either calcareous or siliceous, the former being derived from the shells of foraminiferans, small pelagic mollusks, or flagellate coccolithophorids, and the latter from skeletal material of diatoms and radiolarian protozoans.

Pressure

There is an enormous increase in the pressure of water upon the bodies of animals at great depths, and this may effect the vertical distribution of such species as crabs and mussels (Menzies and wilson 1961). This is not, however, an important limiting factor in the vertical distriburion of animals in general, as internal pressures closely counterbalance external pressures and life is known to exist at the greatest depths. Adjustments of internal pressures are not so rapid, however, to prevent injury in many species that are dredged at great depths and quickly hauled to the surface. Furthermore, individual species have different limits of pressure tolerance.

Temperature and Currents

The temperatures of surface waters vary between the freezing point (1.9°C) in the tro ics. Seasonal variations are small in polar and tropical waters but somewhat greater in the temperate zones.

Temperature varies with depth, more so in the tropics than elsewhere. At 60°N lattitude in the Atlantic Ocean, the mean temperature of the warmest and coldest months beneath the surface is about 10°C. On the equator the temperature beneath the surface is approximately 26°C, at 200m 13°C, at 100m 4.51C, and at 2000m 3.3°C, (Ekman 1943). In temperate regions, seasonal thermocline develops near the surface in summer and is destroyed in autumn and winter when vertical mixing creates a layer of relatively uniform temperature in the upper 20 to 300m.

Currents moving toward the poles from the equator consist of warm water, and currents moving in the opposite direction of cold water. Surface currents make wide circular movements in opposite directions in the northern and southern hemisphere.

Light

Much of the solar radiation is reflected from the surface of absorved in the upper layers. Even in the clearest waters and at maximum radiation; the red, orange and ultraviolet are absorbed in the first 20m. Green, yellow, and blue wavelengths penetrate further, depending on the water colour. When the sun is not at the zenith, light penetration is reduced, and the maximum penetration in the winter at higher latitudes is much less than during the summer.

Salinity

The salinity of sea water varies from place to place depending largely on the amount that it is diluted by the yellow of fresh water from rivers or melting glaciers or the amount that it is concentrated by evaporation. The Red Sea, for instance, has a salinity of 40% (40 g of dry salts in 1000 g of seawater) while in some polar seas the salinity is less than 30%. The average salinity of the oceans as a whole is commonly given as 35%, of which the chloride ion constitutes about 19%, and the sodium ion little over 10%. The various major salts occur nearly everywhere in definite and constant proportions. As one would expect, the pH of seawater is high, averaging about 8, there is some similarity in relative proportions and concentrations of the variations in seawater and in the blood or body fluids of many invertebrate organisms. This may indicate that the sea is the habitat in which living forms first evolved.

The contrast in salinity between seawater (35,000 ppm) and fresh water 15-660 ppm) requires important difference in physiological adjustment of organisms to occupy these two habitats. The problem is one of osmotic regulation (Block 1951). Most marine invertebrates are poikilosmotic in that they are nearly isotonic with seawater; they are highly permeable to water; and gain or lose water according to concentration of the medium. A few marine segmented worms, flatworms, and crabs and all marine fish and mammals have at least some internal osmotic regulation and tend to be homoisomotic. All except the elasmobranch fishes maintain body fluids hypotonic to seawater in various ways. The skin has decreased its permeability to the free movement of water back and forth, the necessary water is obtained by swallowing, surplus salts are secreted outside the body, especially through the gills, and there is a general decreased function or atrophy of watersecreting organs such as the kidneys. The practical absence of insects and amphibians from the sea is largely due to their inability to secrete salts outwardly. The high asmotic concentration found in elasmobranchs is the result of huge quantities of urea retained in the body tissues and fluids.

These salts keep cycling through the ecosystem, but additions to the supply come continually from the land, being washed into the oceans by the rivers. Neritic waters are especially fertile and support a great mass and variety of animal life because of this land drainage and the pattern of water circulation on the continental shelf. Biological productivity decreases progressively from shallow waters over the continental shelf. to deeper waters, to the open ocean, but it also high over offshore banks and in areas of upwelling substantial amounts of nitrogen salts are also swept out of the air by precipitation, and there is nitrogen-fixation by bacteria.

Oxygen

The oxygen supply of seawater comes by diffusion from the air at the surface and from photosynthesis of green plants down to the compensation point. It is continuously used at all depths in respiration of animals and plants and in the decomposition of organic matter.

The oxygen content of seawater (Hedgpeth 1957) is seldom limiting for the occurance of animals, except in the deeper waters of the brackish Black and Caspian Seas, where it is practically absent. Oxygen concentration is especially high on shores where there is splashing of waves. Surface waters of the Atlantic Ocean commonly have 4.5 to 7.5 cc/litre and abyssal regions may run 5 cc/litre. Oxygen is somewhat less abundant in the Pacific and Indian oceans. Oxygen may be reduced to lower concentrations between 100 and 1500 m, because of its use in animal respiration and in decomposition, than at lesser depths, where there is photosynthesis, or at greater depths, where the abundance of animals is greatly diminished.

Marine animals have a variety of mechanism and adoptations for respiration (Flattely and Walton 1922). Greatest difficulties occur in shore animals at low tide when they are exposed to the air, but the need for oxygen at this time is decreased in many forms by curtailment of activity. Some crabs, barnacles, snails, and fish have become almost amphibious in being capable of respiring in air, although at reduced rates as well as in water. Pune mud bottoms may present anaerobic conditions a short distance below the surface, bud mud bottoms mixed with sand contain an abundant and diversified fauna.

PLANKTON

Composition

The plankton of the sea includes a great variety of forms, even more than in fresh water (Biglow 1926, Hardy 1956),

Rotifers, however, are uncommon in marine plankton and cladocerans are much less important.

The nannoplankton consists mostly of protozoans, algae bacteria, fungi, and viruses. The bacteria are largely periphytic, in that they are attached to the surfaces of floating plants, animals, and to particles of detritus. Very few occur freely suspended in the water (Harvey 1955). Bacteria, including nitrifying forms, occur at all depths but are especially abundant in or close to the bottom. They are generally more numerous in the winter than in the summer Nitrogen-fixing organisms are chiefly blue-green algae and possibly yeasts (Oppenheimer 1962).

The green phytoplankton is composed primarily of diatoms, dinoflagellates, and small unarmored flagellates, but several other kinds of algae are present and occasionally important. The dinoflagellates Noctiluca and Ceratium are luminescent and in some regions may give a glow at night to the entire sea. Bioluminescence is not limited to these organisms, however, but occurs also in various forms of bacteria radiolarians, sponges, coelenterates, ctenophores, nemertineans, worms crustaceans, brittlestars, mollusks, balanoglossids, tunicates, and fish (Harvey 1952).

The most important groups of protozoan zoo-plankton, other than the green flagellates which are usually considered with the phytoplankton, are the rhizopod Foraminifera, the actinopod Radiolaria, and the ciliate tintinnids. They may be enormously abundant at times.

Abundance

The actual abundance of plankton varies greatly from place to place and from one season to the next. Smaller species tend to be more numerous than larger ones. The mean annual abundance of diatoms is commonly in the tens of thousands per litre and for shorter periods the year algal blooms may increase the population to hundreds of thousands of cell per litre (Ricketts and Calvin 1948), zooplankton is, however, much less abundant.

Large populations of zooplankton generally follow large populations of phytoplankton and by their grazing maintain the standing crop of phytoplankton at the size suited to prevailing conditions (Nielsen 1958).

The total net zooplankton (cc per unit volume of water) is some 50 times more numerous in the neritic coastal waters of the Atlantic coast of North America than in the Sargasso Sea (Grice and Hart 1962). The abundance of plankton is generally higher in cold than in warm ocean waters correlated with the greater amount of phosphate present in colder water (Harvey 1955). Cold-water plankton tends to be of larger individuals size. There is generally in most taxonomic groups, however, a lower variety of species in cold than in warm waters. Annual productivity is also less in cold waters correlated with fewer generations per year. Productivity of zooplankton varies between 14 and 58 per cent of the net primary productivity.

Nekton

Mollusks, fishes, birds, and mammals make up the nekton of the sea. Mollusks are represented by the squids; fish by the sharks, flying fish herrings, meackerels as many others, including numerous varieties of small species; and mammals, by the seals, porpoises, and whales. The distribution of fish is irregular, but in general they occur more abundantly in neritic waters than in the open ocean. Likewise, they are much more numerous in the epipelagic than in lower strata. Most pelagic fish, except sharks, possess a swimbladder useful for maintaining hydrostatic equilibrium at the depth where there they occur; those fish that lack one are commonly bottom forms (Marshall 1954). In arctic waters fish are less abundant, and mammals relatively more important, than is the cease farther south.

Birds, like many other marine animals, are more numerous in the neritic biochore than in the open ocean. In the ocean far from land occur only penguins, albatrosses, hearwaters, and petrels, and even these species become more common shoreward. Other marine species in nerittic water are tropic-birds, pelicans, gannets, boobies, cormorants, frigate-birds, ducks, gulls, terns skimmeers, auks, and murres. These marine birds may spend many days or weeks feeding and travelling over the water, but all must search out some shore, cliff, or isolated islands on which to nest. Here they sometimes concentrate in enormous numbers during the nesting season because of the limiting number of suitable nesting locations available.

Benthos

Benthos is of much greater variety in marine than in fresh-water habitats. These animals are very abundant in the littoral zone and decrease in numbers with depth until only scattered individuals are found in the deep ocean trenches (Saunders et al 1965). Benthos consists of sessile forms, the sponges, barnacles, mussles, oysters, crinolds, corals, hydroids, bryozoans and some worms; creeping forms, such as crabs, lobsters, certain copepods, amphipods, other crustacoans, many protozoans, snails, echinoderms, some bivalves, and some fishes; and burrowing forms, including must clams, worms, and some crustaceans. Sessile and creeping forms are often grouped as ebifauna are burrowing forms as infauna. Epifauna in the littoral zone decreases in variety towards the poles since it is subjected composition of infauna remains about the same.

In food habsts, different species may be grouped as seston eaters, mostly sessile or semi-sessile forms that capture suspended food particles; sluggish motile forms collecting detritus or food particles from the bottom surface; sluggish forms that extract food particles from bottom material that they swalloos; carnivores, and scavengers.

Oceanic Plankton and Nekton Biome

This biome is characterized by the predominance of organisms possessing life-forms adopted to keep them afloat. Plankton and nekton predominate, although the deep-sea benthos may also be considered as belonging to this biome. Seasonal aspection may bring drastic changes in species composition, especially in plankton. Dominance, in the sense used for terrestrial communities, probably does not exist, except possibly in the Sargasso Sea, where the floating vegetation establishes the habitat. The ecosystem is self-contained, however, since energy is derived from the sun and nutrient material continues to recirculate with little or no dependence on terrestrial resources.

The sargassum community of the Atlantic Ocean is of special interest, the floating Sargassum alga accumulates and is held within a limited area by circular ocean currents. This plant belong to the intertidal zone of the Caribean islands but in turn lose in large amounts along with attached animals during the hurricane season. It continues to grow thereafter, but does not reproduce. The fauna that it contains to a truly littoral one rather than pelagic, but because the algae to cold and ice erosion, but the species accumulates in fresh amounts as fast as old plants die, the animals reproduce and maintain a continued existence far from any shore.

Food Chains

As in fresh-water and terrestrial communities, bacteria in the sea are largely responsible for the final decomposition of excreta and dead bodies to make their essential nutrients available for reabsorption by the green phytoplankton.

Nitrogen and phosphorus are least concentrated near the surface of the ocean, since this is the stratum in which they are most rapidly absorbed by the photoplankton. Excreta and dead organisms sink during the process of decomposition, so nitrogen regeneration is most evident at depths of 500 to 1500 m. The total non-living organic matter, either dissolved or in the form of particulate matter, is generally much larger than the biomass of living organisms at any one time. Organic matter makes up less than one-third of the total particulate matter, and all of it sinks slowly through the water, requiring months or even years to reach the bottom. Much of the particulate organic matter is still undissolved and accumulates on the ocean bottom. Numerous species of invertebrates depend on it for food, especially on the bacteria that it contains. The deep-sea fishes feed on these invertebrates or are carnivorous on other fish. Many of them have very wide mouths, distensible stomachs, and formidable teeth. In addition to these food coactions, it s also likely that many deep-sea fish and larger invertebrates undergo vertical migrations, so that they obtain food by preying on living organims at more moderate depths. Much that is known about the life histories of these deep benthic species has been summarized by Marshall.

Especially fertile regions of the open ocean occur when there is deep mixing of waters by turbulence and upwelling. Vertical water currents bring nutrients up to the surface from intermediate depths where they had accumulated. Prominent regions of more or less permanent upwelling occur around the Antarctic continent, off the coasts of California, Peru, and Somali, and Off the west coasts of both north and south Africa. In many areas, very small phytoplankton and bacteria form the first link in the food chain, which goes next to the microzooplankton before reaching the carnivorous net zooplankton of copepods, euphasiids, and other crustacean forms. Most animals depending on small organisms and organic detritus for food have various filter-feeding mechanisms for straining the material out of the water They do not actively search and catch individual items through directed actions. Invertebrate animals may also be able to absorb some essential salts and dissolved organic compounds to build their skeletal structures and for general metabolism, but there is considerable controversy on this point.

The baleen or whalebone whales (Mysticeti) are toothless but possess large plates in their mouths that strain out the plankton (especially copepods, euphausiaceans, mysidaceans) that they use as food. Only occasionally are small fish as food. Only occasionally are small fish or other invertebrates ingested. Some whales reach tremendous proportions, and the differential in size between these animals and their food is one of the most remarkable in the animal kingdom. Much more common is the feeding on plankton by squids, the young stages of most fishes, and such adult fishesas sardine, an chovy, meuaden, herring, and mackeral. The menhaden is unique in having such fine-mesh gill-rakers that can feed extensively on diatoms, which because of their smaller size cannot be readily secured by other large marine animals.

Balanoid-Gastropod-Thallophyte Biome

This community extends from high-to below low tide levels on rocky shores (Lewis 1964). Benthic animals and attached algal plants are conspicuous and important, the benthos is mostly epifauna, as too hard to permit development of extensive infauna. When the tide is out the organisms are subjected to drying, the occasional inflow of fresh water, higher temperatures, and greater light intensities. Organisms avoid desiccation when the tide is out by variously crawling under stones or thick algal growths, closing thick shells or operculae, retreating into crevices, or secreting a mucous seal. Most organisms are also faced with the pounding action of waves., various holdfast or anchoring devices have developed, and many species protect their more delicate structures with a hard shell. The adoptations for life on the seashore are many and varied (Yonge 1949). The plankton and nekton associated with the benthos include many species not common to the oceanic biome,

Zonation

Vertical zonation of species on rocky shores is usually conspicuous although individual species may extend widely into adjacent areas (Hewatt 1937, Yonge 1949, Stephenson 1949 Southward 1958).

Begining on the landward side there is a supralittoral zone mostly above the action of tides and inhabited as much by land as by marine animals. This is followed seaward by a supmalittoral or Litttorin fringe which is wetted by the highest tides and by the splashing of waves. Because of the presence of either Myxophyceae or lichens, this zone is often discovered; commonly black. The fringe is especially characterized by large numbers of small snails and sometimes isopods.

Next below this fringe is the midlittoral or balanoid zone. It is strictly intertidal being covered and uncovered every day, and is occupied characteristically by acorn barnacles. This zone is often divided into subzones with the barnacles predominate in the upper portion, while polybhaets, colonial hydroids, or other forms are relatively more important in the lower part. The subzonation of algae as often also well marked.

The lowest zone over exposed and then only at extreme low tides, is called the infralittoral fringe. It is a transition area. The entire area between extreme high and low tides, including the mide-littoral fringes, when considered as a unit may be referred to as littoral, eulittoral, or tidal zone to distinguish it from the infralittoral or sublittoral zone that extends from the lowest of low tides to the edge of the continental shelf.

Littoral Zone

Brown algae from thick masses and give protection to those animals that find shelter in or under them. A fauna of copepods, ostracods, water mites, and young littorinids inhabit these seaweeds. In England, the numbers of individual per 100 g of seaweed vary from about 44 on brown algae to over 13,000 on lichens (Colman 1940).

The animal life on rocky shores is varied and luxuriant. Several species of ocorn barnacles, snails, marine impets, marine mussles, goose barnacles, sea anemones, chitons, sponges, hydroids, bryozoans, flatworms, annelids, isopods, crabs, sea urchins, starfishes, tunicates, and insects are present. Total abundance of animals may run into tens of thousands of individuals per square meter.

Sublittoral Zone

Th s community in not subjected to exposure by tides or to the pounding of surf, but is affected considerably by wave action and the complete circulation of water. Animals move around somewhat more freely and there is less need for strong holdfast structures. Most organisms lack physiological tolerance for long exposure to the air and hence differ fundamentally in structure and mores from the community described above.

Laminarias or kelps are the largest of the brown algae and occur commonly in this community. They have root-like holdfast attached to the bottom and their stalks, which are often several meters long, bear leaf-like branches that float of the surface in the larger species. A long list of animals finds shelter and food in the kelp beds and especially in the protection of the holdfasts (Andrews 1945), Polychaet worms are particularly abundant in these holdfasts (Colman 1940). Filamentous red algae (Rhodophyceae) are also prominent

Abundant characteristic animals on the Pacific coast are sea urchins, sea cucumbers, starfishes snails, rock oyster, chitons, limpets scallops, mussels, nudibranches, barhacles, crabs, hermit crabs, hydroids, tunicates, shrims, and various fish. Distribution of fish species correlates strongly with the type of bottom or benthos that is present.

Tidal Pools

Seawater is often retained as depression or pools in the littoral zone and hence organism here are never completely exposed to the air. They are however, subject to high light intensity an increase in the temperature between tides (Klugh 1924). Tidal pools are usually richan both plant and animal life and some species are largely restricted to them. Red algae and kelps prefer the more shaded, cooler pools; the green algae and some of the smaller brown algae prenominate in the well-insolated pools. Animals of both the sublittoral and littoral zones are found here.

PELECYPOD ANNELID BIOME

Habitat

This biome develops depositing sand and mud bottoms in contrast to the biome just described that occurs on eroding rocky shores. There is still a good deal of wave action over sandy bottoms. Fine sand particles shift about almost continuously, and animals have difficulty in preventing their burrows from collapsing. In general, the water over muddy shores is shallower, quieter, and warmer. The mud forms a soft, compact bottom, but is also easily moved or shifted around by storms and wave action. Animal burrows in mud are more permanent. Species tend to segregate depending on the amount of organic matter present. Shores of high mud content may be low in oxygen because of decaying organic matter, so animal populations tend to be largest and most varied in a mixture of mud and sand. Tidal currents are weaker, and change in the level of water is less pronounced on sand and muddy shores than on rocky ones.

Composition and Characteristics

Important plants in this biome are the marine eelgrass, which is seed plant and green algae, particularly the sea-lettuce, which grows in sheets either attached to the substratum or lying fragmented overlarge areas, and Enteromorpha, which grows in tuffs or tangles. Occuring on eelgrass and sea-lettuce may be several kinds of epiphytic algae. These plants form extensive stands and are important to animals for attachment, shelter, and food. Eelgrass was almost eliminated from the Atlantic coast in 1931-32, possibly because of a protozoan disease. This disturbance had a profoundly deleterious effect on the abundance of many animals, including the brant; a bird that depended on it almost exclusively for food, and on scallop and other coastal fisheries. Twenty years later there was evidence that eelgrass was recovering much of its former abundance (Cottam and Munro 1954).

Predominant animals are pelecypods, polychaere worms, particularly Arenicola and Nerels; starfishes, brittle-stars, sea cucumbers, crabs, amphipods, and snails. Populations may run to several thousands of individuals per squre meter. A variety of small fish occur here. Birds include sandpipers, plovers and herons.

The biome is worldwide in distribution, but the characteristic life-fo ms are represented by different species locally. Thus a number of secondary communities (Biociations) may be recognized (Petersen 1914, Jones 1950, Thorson in Hedgpeth 957).

Many of the animal constituents in this biome are burrowing forms The sub-stratum of mud and sand holds considerable water and when the tide is out on exposed flats pelecypods, worms and other animal constituents retract their fleshy argans into their burrows shells and remain in a water saturated environment (Hesse, Allee and Schmidt 1951). They are thus not exposed to the atmosphere with changes in the tide Furthermore, most forms are generally tolerant of low oxygen and high carbon dioxide concentrations. In order to maintain respiration when retracted in their underground burrows these animals have long siphones, sometimes longer than their bodies, or long tubes or canals that extent to the surface. Through these they maintain a circulation of water, often by means of special pumping organs.

Coral Reef Biome

Coral reef are formed by the accumulation of the calcareous skeletons of myriads of organisms. They extend from the sea bottom at depths of 46 m or rarely 74 m. to slightly above low-tide level. The best formation of coral reefs is confined to warm waters above 10 C, although individual species may extend into colder regions (Vaughan 1919, Wells in Hedgpeth 1957).

Predominant organisms involved are commonly the anthozoan stony corals and organ corals and the hydrozoan millipórids. Some reefs however are formed principally by foraminifera and still other by calcareous algae. All massive coral structures employ calcareous algae as cement. These algae not only thrive in the pounding surf on the windward side of the reef, but by their growth are able to repair damage to the reef caused by storms. Most typical reef-building animals are colonial and of shapes varying from closely compact, globose, or encrusting, to loosely branched or dendritic, depending in part on their exposure to wave action. Each polyp in a colo᷊y secretes its own calcareous skeleton and when it dies the next generation builds on top of the old so that the accumulation of a lime structure is fairly rapid (Yonge 1963).

The bright yellow or red colors of corals near enough to ther water's surface for adequate light penentration are the result of algae, the zooxanthellae, which are either embedded in the body wall or free in the internal cavities. In addition, there are bands of green filamentous algae growing to a ‹ epth of 2 or 3 cm in the pores of the insert coral skeleton that may have a biomass 16 times that of the zooxanthelale. In their photosynthesis these algae may absorb dioxide and nutrients derived from animals or the coral reef and liberate oxygen of value to the animals (Odum and Odum 1965). Perhaps because of this symbiotic relationship which requires solar radiation, living corals are largely confined to the upper, shallower waters. Coral animals actively ingest zooplankton, but apparently not phytoplankton, from the surrounding water.

Succession to Land

The three great biocycles-ocean, freshwater, and land-come into contact with each other around the margins of the seas. The change in the physical nature of the habitat from salt water to fresh water is a drastic one, but not more drastic than the change from salt water to land : The transition of animal and plant life is brupt, and a zonation or physiographic succession of communities can be recognized. This transition from the ocean to fresh water and from the ocean to land as we see it today is of special interest since it parallels the probable evolution and dispersal of life in past ages.

Life is generally believed to have originated in the littoral region. Apparently no great group (phyla) of animals originated except in the ocean. The routes by which animals probably left the ocean and reached fresh-water and land have been various. Some animals probably migrated directly across sea branches; others probably ascended rivers, passed through marshes and swamps, or burrowed through soil. Some animals were transferred from the ocean by land elevations which isolated them in bodies of water which gradually become fresh... Imigration from the sea did not take place at one time. It has occurred many times in the past and is slowly progressing on many shores today The most successful animal colonizers of the land have been :

1 The arthropods, which have in many cases developed book-lungs or tracheae for breathing air;

2. The vertebrates, with lungs and dry skins; and

3. The snails with slime and spirally coiled shells to prevent desiccation There are at present many examples of animals which are in the midst of their transformation from marine to fresh-water animals, or from marine of fresh-water into land animals. Not only have plants and animals emigrated from sea to land, but there are countless instances when migrations have taken and are taking place in the opposite direction. Grasses, insects, reptiles, birds and mammals have left the land for the sea..Fishes began in fresh water, but new range through the ocean at all depths.

On rocky shores and cliffs there is a splash or supralittoral zone above high tide level. Green algae occur here and scattered individuals of marine snails, acorn barnacles, limpets, amphipod sandfleas and flatworms as well as insects, especially. Diptera and other forms that come from the land. Above the influence of splashing, the rocks may be covered with lichens and mosses, representing the initial stages in the terrestrial rock sere. However, salt spray is often blown inland a considerable distance to affect conspicuously the development of normal terrestrial vegetation and its accompanying animal life. Cliffs along the ocean, as well as sandy beaches and islands, are favourite nesting places for large numbers of pelagic birds.

Succession to Fresh Water

Where rivers flow into the ocean on low coastal plains and there are extensive embayments or estuaries as long the Atlantic Coast, there is very gradual change from salt water to brakish water (salinity : 0.5 30%0) to entirely fresh water. This habitat gradient fluctuates back and forth with the tides. Since fresh water is less dense and often warmer, it flows over the top of the salt water, with the result that strata with different physical characteristics are formed and these different strata are inhabited by different kinds of fish and other organisms, influx of fresh water is one of the principal sources of dissolved nutrients, and the gradient of salinity is dependent upon continuous inflow of fresh water. Any reduction of inflow, as by damming of streams, will allow salt water to penetrate further inland.

APPLIED ECOLOGY

Food Production

Although 71 per cent of the earth's surface is occupied by oceans and only 29 per cent by land, nearly all of the food and raw materials used by man is derived from the latter. This is in spite of the fact that agricultural soil is only a few inches thick and must be cultivated, protected from erosion, and fertilized, while the ocean with its chemical fertility, its photosynthetic production of basic plant food, and its great fisheries is an entirely natural resource ready to be harvested. However, fish harvesting in the ocean has definite limits determined by photoplankton primary productivity. It is estimated that maximum sustained yield to man is of the order of 100-150 million metric tons per year and in 1967, 6000,000 tons were being harvested. The ocean is relatively infertile, yet it covers most of the area; much more productive are the coastal areas, particularly the areas of upwelling but these are of comparatively small size. A large part of the production is taken by predators other than man and part of it must be left to maintain the breeding stock. The possibility of the ocean serving as an abundant future source of food for man is probably exaggerated. Seafood is important mostly as a supplementary source of protein. Man must look to other resource for his main energy requirements.

The plankton of the sea represents a possible food supply for man (Davis 1955). Its energy value is approximately 4 kcal/g dry weight, and it is more or less palatable (Clarke and Bishop 1948). However the expense involved in filtering and concentrating it is too great to be economical; poisonous species sometime occur, it is not easily digested and assimilated, and consequently it has not as yet proved to be a feasible diet. The energy value of the plankton is used by man at the present time primarily as it is transferred into higher links of the food chain. Aside from the fishes, the chief marine organisms used as food are the oysters and other mollusks, shrimp, crabs, lobsters, and sea turtles. These are mostly animals of the continental shelf and estuaries.

There are a number of problems in the use and conservation of marine organisms. Natural beds of American oysters on the Atlantic coast have nearly all been exhausted through over-fishing and pollution. Because of heavy erosion of the land, silt deposition has become excessive in most of the bays and estuaries and there is increasing difficulty for oyster spat to find clean hard surfaces on which to set Surfaces that are loose or covered with silt are not suitable since the spat is very small and easily smothered. A common practice is to return to suitable areas all shells of mollusks removed or to introduce other suitable hard objects to furnish the necessary substratum for oyster setting. Control of erosion over the watershed would greatly alleviate the problem. The trend is increasing to lease suitable areas of water and to farm oysters in the manner of an agricultural crop.

In spite of their position at or near the top of the food chain, the greatest utilizable food resource of the sea is its fishes.

Ocean "farming" is largely limited to estuaries and salt mashes. Yield in the Orient occasionally reach or even surpass 1000 kg/hectare-yr (880 lb/acre-yr). Principal species are milkfish, grey mullets, eels, salmon, sea trout, tilapia, oysters, and shirmp. Conversion of marsh into ponds where water level, fertility, and cropping can be controlled has met with varying success in different parts of the world, but may be expected to expand as fish harvesting in the ocean itself approaches its yield limit.

Summary

Geographic distribution of marine organisms depends on their responses to current temperature, and physical barriers; their local distribution is affected by waves and tides, type of bottom, salinity, and depth. Major division of the marine biocycle are pelagic (open water) and benthic bottom). Major communities recognized are the oceanic plankton and nekton biome in the open sea, the balanoid-gastropod-thallophyte biome on rocky shores, the pelecypod-annelid biome on sand and mud bottoms, and the coral reef biome.

Organisms making up the oceanic biome are widely distributed around the world but may be divided into warm-water, arctic, and antarctic faunas. Coral reefs are found only in the tropics. The two biomes on the continental shelf subdivide into warm-water, temperate, arctic, and antarctic faunas and into more restricted regions and subregions. The warm-water faunas are richest in species, especially in the Indo-Malayan and West Indian subregions.

Marine plankton include a greater variety of forms than does fresh-water plankton, although rotifers are nearly absent; cladocerans, less important. They possess various unique mechanisms for flotation. Although abundance varies greatly from place to place and from season to season, plankton is generally much more numerous in neritic coastal waters than in the open sea. Diel movements between the surface at night and greater depths during the day are pronounced.

Mollusks (squids), fishes, birds, and mammals constitute the nekton. The taxonomic composition of the fish fauna varies with depth. Bioluminescence is exceptionally well developed among deep-sea nekton and benthos.

Benthos includes a great variety of sessile, creeping, and burrowing forms. It is very abundant in the littoral zone, and decreases in numbers with depth, although individuals are found in the deepest ocean trenches. There is considerable difference in the life-form and species composition of benthos occurring on rocky shores and on sand and muddy ones. Zonation of species is more prominent on rocky than on depositing shores. Succession and dominance occurs in some situations, but is less important than in terrestrial communities. Coral reefs have many special features, but are now exposed in the Pacific and Indian Oceans to damage by a species of starfish.

Food chains in the sea are similar to those in fresh water but the number of links and species composition varies in different communities. Productivity is especially high in regions where upwelling and turbulence bring nutrients from deeper levels up to the surface.

The three great biocycles of ocean, fresh-water, and land come into contact around the margins of the seas. The succession from the ocean to land is abrupt and from the ocean to fresh water through estuaries only slightly less so.

Effective use of the ocean resources as food for man is largely limited to the continental shelf, regions of upwelling, estuaries, and coastal marshes. Plankton is abundant and has high productivity in these regions but is likely to be useful to man only as its nutrient material and energy are transferred with considerable loss into the higher food chain links of mollusks, larger anthropods, fishes, and other vertebrates. Coastal regions need protection and management for good production and all waters kept clean from pollution of all sorts.

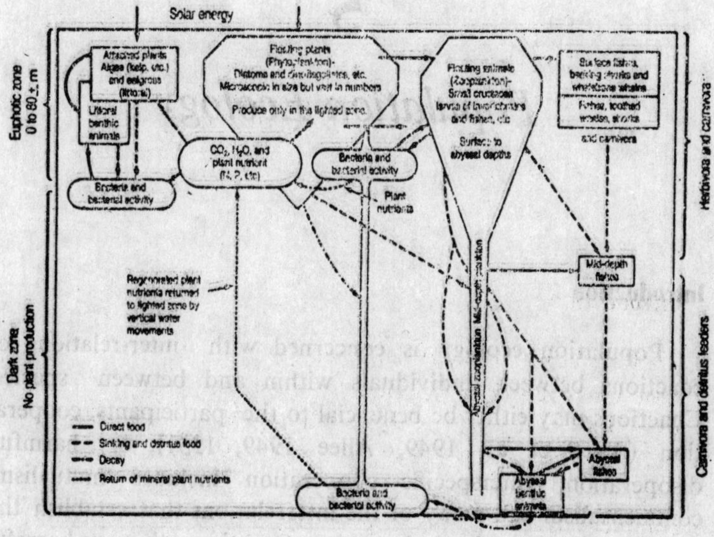
Solar energy
Euphotic zone
Dark zone
No plant production
Attached plants
Algae (kelp, etc.)
and eelgrass
(littoral)
Littoral
benthic
animals
Bacteria and
bacterial activity
Floating plants
Produce only in the lighted zone.
Surface to
abyssal depths
and carnivora
Mid-depth
fishes
Abyssal
fishes
Herbivora and carnivora
Carnivora and detritus feeders
Regenerated plant
to lighted zone by
vertical water
movements
Direct food
Sinking and detritus food
Decay
Return of mineral plant nutrients
Bacteria and
bacterial activity
Abyssal
benthic

5

Population Ecology

Introduction

Population ecology is concerned with inter-relations or coactions between individuals within and between species. Coactions may either be beneficial to the participants, cooperation (Allee et al. 1949, Allee 1949, 1951), or harmful, disoperation. Interspecific cooperation includes mutualism, commensalism and many of the interrelations that establish the community as a dynamic unit. Coactions that are harmful to at least one of the participants, include parasitism, predation, and competition. Many of these coactions act to regulate density or, along with climate, cause fluctuations or even more drastic changes in number of individuals from time to time. Established behavioural interrelations between individuals, characteristic of a species, are its ethics and as such are found in animals as well as in man.

INTRA AND INTER SPECIFIC COACTIONS BETWEEN INDIVIDUALS

Intraspecific Cooperation

An early manifestation of cooperation in the evolution of animals is the grouping of free-living protozoans to form colonies, and the further development of such colonies into multi-cellular metazoans that thereafter behave and respond as unit organisms. Whether the first gathering of protozoan cells to form colonies developed for better protection from some predator or environmental condition, improved utilizalion of food supplies, or more efficient reproduction, it is impossible to say. The colonial form, however, must have had survival value to persist.

Aggregations

All organisms react on their habitat in one way or another, and when they occur in numbers these reactions produce a conspicuous effect. Water conditioning occurs when physical or chemical changes occur as the result of organisms living in it. Compared with unconditioned water, these changes may have either a harmful or a beneficial effect on organisms introduced into the water after the original organisms have been removed. Water is said to be homotypically conditioned when the changes were previously produced by individual of the same species as being studied and heterotypically conditioned when the changes were produced by a different species.

Mutualism

Mutualism is an association between two or more species in which all derive benefit in feeding or in some other way. The term symbiosis has often been applied to this relationship, but symbiosis properly refers to the intimate association of two or more dissimilar organisms, regardless of benefits or the lack of them, and hence includes mutualism, commensalism, and parasitism.

Mutualism, as is true also with commensalism, and parasitism, may be facultative, when the species involved are capable of existence independent of one another, or obligate, when the relationship is imperative to the existence of one or both species. Considerable study and experimentation is sometimes required to decide whether a particular relationship is facultative or obligate, or even whether it is truly mutualistic.

Mutualism in plants is demonstrated in the associations of fungi and algae to form lichens, of nitrogen fixing bacteria with the roots of legumes, and of fungal mycorhizae with the roots of many flowering plants.

There are many intimate relations between plants and animals. Mutualism is suspected in the presence of photosynthetic algal cells in the protective ectoderm of green hydra, and those associated with turbellarians, mollusks, annelids, bryozoans, rotifers, protozoans, and the egg capsules of salamanders. The álgae give off oxygen, benefiting the animals, which in turn supply carbon dioxide and nitrogen to the plants. The thick growth of algae often found on the carapace of the aquatic turtles is important mostly as camouflage for the turtles. Certain beetles, ants and termites cultivate fungi for food. Bacteria in the caeca and intestine of herbivorous birds and mammals aid in the digestion of cellulose. The cross-pollination of flowers by the agency of insects and birds seeking nectar and pollen is of such great importance that many structural adaptations in both plants and animals fit the one to the other to ensure the success of the function.

Commensalism

Commensalism defines the coaction in which two or more species are mutually associated in activities centering on food and one species, at least, derives benefit from the association while the other associates are neither benefited nor harmed. It is often difficult to establish definitely the nature of the relations between species; and phenomena considered at one time to be commensalism have been later found to be parasitism or mutualism. The concept of commensalism has been broadened, in recent years, to apply to coactions other than those centering on food; cover support, protection, and locomotion are now frequently included (Baer 1951).

The remora fish are remarkable for having the spinous dorsal fin modified to form a sucking disk on top of the head by means of which they become attached to the body of the shark, swordfish, funny, barracuda; or sea turtle. They are of small size and are not burdensome to the host. The host benefits the remora, however, for when the host feeds, the scraps of food floating back are swept up by the remora.

Parasitism

Parasitism is the relation between two individuals wherein the parasite receives benefit at the expense of the host; parasitism is therefore a form of disoperation. Parasitism is mainly a food coaction, but the parasite derives shelter and protection from the host, as well. A parasite does not ordinarily kill its host, at least not until the parasite has completed its reproductive cycle. Where the parasite kill its host immediately on infecting it, the parasite would be unable to reproduce and would quickly become extinct. The balance between parasite and host is upset if the host produces antibodies or other substances which hamper normal development of the parasite. In general, the parasite derives benefit from the relation while the host suffers harm, but tolerable harm.

Social parasitism describes the exploitation of one species by another, for various advantages. Old world cuckoos and the brown-headed cowbird of North America do not build nests of their own; rather, they deposit their eggs in the nests of other species, abandoning eggs and young to the care of foster parents.

Parasitoidism and Predation

Some Diptera and Hymenoptera deposit their eggs in the immature stages of other insects; the larvae on hatching feed on the host until they are fully grown. The relation of the larva to its host is frequently described as parasitism. It is fundamentally different from parasitism in that the host generally dies of the larva depredations before the larva emerges, but the larva lives inspite of the death of the host. The relationship resembles that of predator to prey, except that, unlike the true predator the larva lives within the body of its prey and kills it slowly as it feeds. Such larvae are called as parasitoids.

Predation is a form of disoperation, at least in point of immediate effects, since one animal kills another for food. Predation is important in community dynamics in so many ways.

Competition

Competition is the more or less active demand in excess of the immediate supply of material or condition exerted by two or more organisms (Clements and Shelford 1939; p. 159). The materials and conditions sought by animals include food, space, cover and mates. When these materials are in more than adequate supply for the demands of those organisms seeking them, competition does not occur; when they are inadequate to satisfy the needs of all the organisms seeking them, the weakest, least adopted, or least aggressive individuals are forced to do without, or go elsewhere. Competition may result in death for some competitors, but this is from fighting or being deprived of food or space rather than being killed for food as in predation, or by disease as in extreme parasitism.

Competition may be either direct or indirect. It is direct where there is active antagonism, struggle or combat between individuals; indirect, when one individual or species monopolizes a resource or renders a habitat unfavourable to the establishment or other organism having similar requirements. Direct competition, or interference, is evident in the fighting of bull seals for larger harems and of grouse for a better position in the social hierarchy; in chasing and color displays (a sort of saberattling) by fish and birds for defense of territories; in the singing and calling of birds, some mammals, and frogs as bids for mates; and in the excretion of chemicals that effect the behaviour or health of other organisms (allelo-chemistry).

Competition is usually keenest between individuals of the same species, intraspecific competition, because they have identical requirements for food, mates and so on. And because they are more nearly equal in their structural, functional and behavioural adaptations. Interspecific competition occurs where different species require in common at least some materials or conditions. The severity of competition occurs where different species require in common at least some materials or conditions. The severity of competition depends on the extent of similarity or overlap in the requirements of different individuals and the shortage of the supply in the habitat. It is generally the case that the more unlike the kinds of competing organisms, the less intense the competition.

Allelochemistry

Included here are the coactions whereby chemicals secreted by one organism affects the growth, health, or behaviour of other organisms. Allelopathy is produced in plants when toxins are liberated that inhibit seedling growth in the vicinity. This may affect succession of plant species, especially important in the early stages. Some pioneer species in the abandoned field sere produce substances inhibitory to nitrogen fixing and nitrifying bacteria. This retards invasion of other species that require higher nitrogen concentration in the soil. Volatile inhibitors are generally more prevalent than water-soluble ones and relatively more prevalent in arid than humid climates. Antibiotics produced by bacteria, fungi, actinomycetes, and lichens are widespread in nature and may be one of the reasons why bacteria pathogenic to man cannot multiply well in soils. A number of antibiotics, such as penicillin, have been used extensively in human medicine. The use of allelochemic coactions in agriculture has possibilities but has yet to be exploited.

Among animals, overcrowding of tadpoles in culture dishes is associated with the occurrence of peculiar round vacuolated cells in the intestinal tract and feces that appear responsible for curtailment of further growth. Under laboratory conditions, killer stocks of Paramecium aurelia produce a toxin, paramecin, at the rate of one unit particle per 5 hours. One unit particle is enough to kill one individual of sensitive stock of the same species as well as being lethal to other species of Paramecium. Conditioning that becomes unfavourable homotypically may sometimes be favourable, or at least tolerable, heterotypically. Thus in protozoan infusion there is a microsere of one species succeeding another.

Allelochemic effects are of great variety in both plants and animals; repellants, escape substances, suppressants, venoms, inductants, counteractants, attractants, signals, stimulants, autotoxins, antoinhibitors, and so on. Pheromones are chemical messages between members of a species especially important in reproductive behaviour, social regulation and recognition, alarm and defense, territory and trail marking, food location, and so on. Many of these effects are beneficial to the individual; others serve for competitive purposes.

Social Hierarchies

When groups of individuals of certain animal species are confined to limited areas, frequent fights or pecking of one another occur. By way of these encounters, the more aggressive and successful individuals establish a hegemony to which the more submissive individuals acquiesce. A social hierarchy is thus established; the phenomenon was first clearly described for the domestic fowl.

The peck-order in the domestic fowl is a linear one. Close observation of marked individuals showed that, in a flock of 13 birds, one bird became the supreme despot of the whole flock; another bird was submissive to the first but despotic over the remaining 11; and so it went on down to the last bird, which had the right to peck none but was pecked by all. This type of social aggressiveness or despotismis called peck-right. In practice, certain individuals establish the right to peck others and not get pecked back. In the middle of a series, the order is sometimes less fixed, and reversals or triangles occasionally occur. Although most easly demonstrated in the crowded conditions of captivity, peck-right has also been observed to obtain under free natural conditions. The peck-right type of social hierarchy has been found to occur in various degrees of perfection in several other species of birds, in several species of mammals and fish, and in a few lizards, frogs and toads, crayfish, and insects.

Possession of the following characteristics usually gives an individual at least some advantage in gaining a high position in the despotic order : strength, good health, maturity, relatively large size, hegemony over own territory, responsibility of acting to protect young, accompaniment by members of his own group when meeting a stranger, male over female during the non-breeding season, female mated with a strong male, the hormone testosterone, and innate aggressiveness. A high position in the social order is advantageous to the individual, as it gives him priority over food, mates, territory, and other resources of the habitat and is sometimes, but not always, correlated with leadership in the group.

In some species the social hierarchy is not as overt as that we have described. In peck dominance the individual that is usually subordinate is successful in a certain number of conflicts. Position in the despotic otder is a function of ratios of success in continuing conflicts rather than of the results of the initial contact an individual has with each member of the group. An individual can occupy any position in the hierarchy as long as he is able to maintain that position against all challengers. The individual who has successfully challenged a higher position from a lower moves up to the higher. Plainly the positions in the hierarchy are fixed in order, but occupancy of those positions is fluid. A still more fluid form of social aggressiveness is supersedence, in which a successfully challenging individual usurps the position of another individual, momentarily possessing special advantages in the presence of food or some other thing. This type of relation has been described for the golden-crowned sparrow and may likely be found in many other species.

There have been few studies of social despotism as an interspecific phenomenon. The range of aggressiveness between individuals within any species is so wide that strong individuals of one specles may be despotic over weak individuals of another even though the majority of individuals in the first species are submissive. However, the sharp tailed grouse is usually dominant over the ring-necked pheasant, and the latter is usually dominant over the prairie chicken. The manner in which different species fit into a social hierarchy may be a key to structure and organization of communities.

Reproductivity and Population Structure

The rate at which a species reproduces, or its reproductivity, and the frequency of its population turnover can affect the speed with which it occupies new areas, becomes adapted to new niches, or evolves into new races In order to analyze the population dynamics of a species, it is necessary to know its life history. This involves the stages in its life cycle, mortality rates of each stage, longevity, sex and age ratios, age at which individuals become sexually mature, fecundity, factors causing mortality, and so forth. The proportion of different ages and sexes gives the population a definite structure. All these essential data may be conveniently summarized in the form of life tables.

Fecundity

Species vary greatly in the characteristic number of generations, broods, or litters produced per year, and in the sizes of them. Protozoans often divide so rapidly that they produce a new generation every few hours. Plankton organisms, less fecund, may produce a new generation every few days. Many vertebrates breed butonce a year, some large animals only once every two or three years. Several species of small birds and mammals have two or more broods per year. The female woodland white-footed mouse in Michigan may produce three litters between early April and early June, and two more between middle August and early October. Under favourable environmental conditions, rodents may continue to breed throughout the winter, so that their reproductive potential is enormous.

Parental Care

The number of eggs or young produced per litter is correlated inversely with the amount of attention that they require. When parental care is altogether lacking, invertebrates may lay 1,000 to 500,000,000 eggs at one maturation; where there is some protection afforded by brood puches 100 to 1,000 eggs may be laid; with a high degree of broad protection, 1 to 10 or more eggs may be laid. Mammals seldom have more than a dozen young in a single liter and, in larger species, usually only one. Characteristic clutch size among birds varies from 1 to 15; rarely, 20.

Death Rate

Death rates vary among species and are correlated with rates of reproduction. The death rate of a species is influenced by a number of factors, but of fundamental importance is the number of young that are born in relation to the carrying capacity of the habitat. When more young are born than the habitat can support, the surplus must either die or leave the area. When populations are stabilized at a constant level, the death rate must fluctuate with the birth rate. Evolutionary adaptation tends to lower the frequency at which the population replaces itself and to raise reproduction to the highest rate compatible with the energy resources both of the species and of the habitat.

Survival of Young

Success in raising young depends not only on the ability of the adults to care for the young, but also on the vitality of the embryo and on the chance destruction of nests, eggs, or young by storms wind, floods, predators, accidents, and desertion of the parents. Considerable data are available in this connection with birds.

Nest failures in birds are most frequent early in the nesting proceeds : 2.4 per cent per day during nest-building, 2.2 per cent per day during incubation, and 0 5 per cent per day while the young are in the nest. Location of the nest is a factor in the successful raising of young.

Life Tables

Species differ widely in the number of young produced each year, in the average age to which they live, and in their average rate of mortality. When sufficient facts about a species are known, a life table that tabulates the vital statistics of mortality and life expectancy for each age group in the population may be formulated (Pearl 1923). Age is usually represented by the subscript index and is some convenient fraction of a species mean life span such as a year or stage of development. The life table is set up on the basis of an initial cohort of 100, 1000 or 100,000 individuals; and the number living to the beginning of each successive age interval is symbolized as I_x.

Life tables are also useful for computing the average longevity of a population, for showing the age composition of a population, for indicating critical stages in the life cycle at which mortality is high, for showing differences between species. for showing the success of the same species in different biotopes, for furnishing information of value in game and fish exploitation (yield), and in control of pests.

Information for constructing life tables may be obtained from a knowledge of age at death of a random but adequate sample of the population; information on the age ratios of the living population, provided it is stablized; or from data on a single cohort, adequately identified, followed throughout its life span.

Sex Ratio and Mating Behaviour

The primary sex ratio at the time the eggs are fertilized should be approximately 50 males : 50 females in most species, although it has seldom been measured. This ratio may be displaced in one direction or the other by differential mortality of the two sexes during the period of growth, become manifest in the secondary sex ratio at the time of hatching or birth, and even more pronounced in the tertiary sex ratio of the adults (Mayr 1939). Recognition of sex and age in living animals is often difficult, although criteria have been worked out for many species.

Breeding Age

The age at which young animals first attain the ability for reproduction affects the reproductive capacity and rate of growth of populations. Planktonic entomostracans are sexually mature in a few days; insects, often in a few weeks. Among birds, a tropical sparrow is khown to reach full reproductive level in 6 to 8 months. Small non-tropical song birds commonly nest during spring and summer of the year followi g that in which they hatched but banding of nestling house wrens indicated that 12 to 18 per cent failed to do so until the second or third year. Upland game birds probably nest as yearlings; geese and wild turkeys do not nest until they are 2 years old; common terns, commonly only after 3 years to reach sexual maturity; turtles much longer.

Non-Breeding Populations

Although the breeding population of a particular mammal, bird, or other animal is only fraction of the total population of a species concerned with its reproductivity, there is often present in an area a substantial, though inconspicuous, non-breeding population that must be considered in any understanding of community dynamics studies have shown that non-breeding population especially of birds, consists principally of young that have been slow to reach sexual maturity, of surplus individuals of either sex in monogamous species, and of adults which, for one reason or another, have lacked reproductive vigour or have been unsuccessful in establishing proper breeding conditions.

Longevity and Mortality Rate

When protected in capacity, animals are capable of living surprisingly long periods. Definite physiological limits of life are characteristic of each species and are occasionally realized under natural conditions, but invariably the potential longevity of a species in many times greater than the mean longevity actually attained by wild populations. Protozoa and some coelenterates and flatworms are potentially immortal when they reproduce by fission.

Age Ratios

The life table gives the number or percentage of individuals in a broad or litter surviving to the next age level. From such data, as well as by occasional direct observation, it is possible to determine the age structure of a population at any one time, the percentage of each age class in populations of adults having three different mean survival rates in all age classes. It is at once apparent that the number of age classes in a population is greater when survival rates are high than when they are low. It is also evident that there is less difference between number of individuals in succeeding age classes when survival rate is high than when it is low.

The exact age of the sexually mature adult is usually at best difficult to determine unless one can band or mark, the young when they first appear, or unless there are growth rings, such as in the scales and otoliths of fish and in the shells of clams, or other criteria of age that can be used. Immature animals are often distinguishable from adults, so that adult young ratios are usually obtainable and often yield important information. The proportion of immatures to adults is highest at the end of the breeding season, and then usually declines until the beginning of the next period of reproduction, because of the higher mortality rate of the young compared with that of adults.

Age ratio is of practical value in wild-life management. A low ratio of immatures to adults indicates a poor reproductive season and should caution against excessive take or yield, as the population is declining.

Adaptation to Biotope

In the stabilized population of any species, whatever the number of eggs or young produced per pair of adults, the average number of offspring reaching reproductive status can never be greater than two in sexual forms, which is the number required to replace the parents on their death. With each new generation there is, therefore, a population turnover, with newly born individuals replacing the adults that die. In a stabllized population, the rate of increase of population through the course of several productive cycles must equal the death rate, so that the value of one factor is also a measure of the other. Either factor is indicative both of the rate of population turnover and of the intensity of environmental resistance.

There is a general inverse relation between growth rate and longevity, T. However, if growth rate were dependent only on the longevity of the species, then rmT would be a constant. Obviously this is not true. It appears that different growth rates may correlate with various intensities of environmental resistance in the different habitats occupied by different species. If there were a habitat offering no environmental resistance, and all offspring therefore survived, then a female would need to produce only one female offspring to replace herself when she died. This would be the net production rate. Actually, the number of female offspring that must be produced to offset mortality caused by the environment is always more than one, attesting the rigor of the nature environment in spite of the species adaptations for life in it. Reproductivity is the rate at which a species reproduces. The number of offspring raised to maturity per unit of time is generally characteristic of a species and varies with fecundity and survival of the young. Fecundity depends upon the morphological and physiological capacities of the species, the amount of parental care that the offspring receives, and weather conditions. Death rates correlate directly with the number of young produced. Reproduction cannot be considered successful unless the young reach sexual maturity.

Life tables tabulate, in condensed form, the vital statistics of survival and mortality by time intervals. They provide essential data for calculating longevity and age composition of populations. Survivorship curves show three characteristic survival patterns, but most populations exhibit a relatively high death rate early in life and a lower, more constant, death rate thereafter.

Sex ratios are often correlated with mating behaviour. The age at which full reproductive maturity is attained varies widely between species. Young birds, surplus adults of either sex, and birds unsuccessful in establishing breeding relations sometimes constitute a relatively large non breeding population in addition to the more conspicuous breeding one. Perhaps this is true also for other animals, but evidence is scanty.

Ratios of young animals to adults often indicate whether a population is expanding, contracting or is stabilized. In stabilized populations the number of offspring reaching reproductive maturity can never be greater or less than the number of adults themselves. The number of young that must be produced to permit such a population turnover gives a measure of the rigor of the environment, and how well adapted a species is to its niche.

Irruptions, Catastrophes, Cycles and the Population of Man

Many theories have been advanced as to the immediate cause of population cycles; diseases, conditions between prey and predator, depletion of food supplies in quantity or quality, physiological stress and natural selection. Population may increase or decrease through a period of time. They may suddenly, except when conditions become favourable or decline precipitatively with unfavourable weather or failure of food supply or they vary cyclically. In this part of our lesson, we are having a discussion of these factors.

Abundance may change continuously and progressively in one direction over a long period of time, or variations in abundance may take the form of irruptions, catastrophes, or cycles. An understanding of how and why such changes in abundance occur is of considerable academic interest, and is of the utmost importance for the economic management of fish and game, preservation of wildlife and in animal husbandry, agriculture and forestry.

Minor fluctuations of less than 2 ; 1 or 3 : 1 are often the result of sampling errors in estimating the true size of the population. When the ratio of population sizes from one period to another is greater than can be explained by errors of sampling, the fluctuations have meaning for which we should know the causes. Population ratios from one year to another are commonly of the order of 10 : 1, 100 : 1 or, in insects, up to 10,000 : 1 or more.

Irruption, Outbreaks, Plagues

The phenomenon of a population suddenly exploding to supersaturate an area is called an irruption, outbreak or plague. These terms are considered here to be synonymous and to represent the time when an animal is abundant or injurious enough over an appreciable area to be noticed and recorded by untrained observers (Carpenter 1940b). The number of rodents may be in the hundreds or thousands per hectare, of insects in the millions. Outbreaks are known to have occurred since the beginning of recorded history of Europe. Africa and North America, especially in insects and rodents, plagues of European meadow voles were recorded 18 times in France between 1792 and 1931.

The cause and control of plagues or outbreaks have concerned man since civilization begin outbreaks are commonly associated either with favourable weather and food conditions so that survival and reproduction is increased, or with a disturbance in predator-prey relations. Rapid increase in population size may thus occur with almost any kind of animal in any habitat. Irruptions of the bean clam occurred several times between 1894 and 1955 in the intertidal zone at La Jolla, California. The abrupt decline of the last outbreak in 1951-52 was the result of an epizootic associated with a minute unicellular organism of uncertain identity, found in the tissues of the clam.

Catastrophes

Catastrophes occur at more or less widely spaced intervals and bring marked depressions in the population level of a species. Information from observers indicated that low populations were widespread in eastern North America. both in this species and in many other song and game birds during most of these years. There is considerable evidence that these conspicuous variations occur in abundance.

Among mammals, fluctuations in the population of the common hare in Denmark have been correlated with the varying effects of summer rainfall, spring temperatures, and the number of days of frost during the winter.

Catastrophes may occur with practically any type of animal life. The severe winter of 1917-18, for instance, produced a marked reduction in the number of many species of marine invertebrates in the region of Woods Hole, Mass. In seeking correlations between catastrophes and weather or other environmental conditions, one needs to determine the critical period in its life cycle during which the species is most vulnerable, and the weather conditions coincident with that period that exert greatest effect.

Cycles

Populations are said to be cyclic when they alternatively irrupt and subside in a more or less uniform manner between high and low levels of density.

Reality of Cycles

To designate fluctuations as cyclic implies considerable regularity for them. What ecologists call cycles are really oscillations because both phase and amplitude are inconstant. The variability that is evident, however, especially in phase, is less than is to be expected by chance, and reasonably accurate predictions can be made of the course of future variations in population size.

The short-term cycle is commonly 3, 4 or 5 years long, although it may be as short as 2 years, or as long as 6 years. The coefficient of variation standard deviation divided by the mean, for different species having the short cycle, varies from 30 to 50 per cent, and is of the same order of magnitude for the longer cycle. It is of interest that by drawing numbered cards from a well-shuffled deck or rolling dice or plotting random numbers short and long cycles may be obtained of about the same relative lengths and variation coefficients as animal population cycles.

There has been considerable controversy about the true nature of cycles that is, whether they are caused by a master factor or factors or by the random impact of a large number of factors acting more or less independently of each other. Before oscillations in the size of natural populations are considered cycles, they should be tested statistically for randomness. Even then, the cause of cyclic populations can be expected always to contain some random components inextricably mixed with any systematic ones. The reality of cycles may also be tested by the amount of synchrony that they exhibit in different localities and on different species. The presence of such synchrony would indicate the presence of broad coordinating influences or mechanisms rather than haphazard effects of many random factors.

Intrinsic Causes

Intrinsic factor affecting animal cycles are those inherent in the populations themselves and in the interrelations of different species. They are primarily biotic in nature. A number of general the ories of possible intrinsic mechanisms have been found inadequate, and still other theories have limited application. However, disease, predation, food and genetical or physiological explanations merit consideration.

Extrinsic Causes

Extrinsic factors that may affect cycles are largely abiotic and external to the populations, such as the weather. Weather is affected by variations in solar radiation.

Solar radiation may also affect animals and plants in other ways than through the weather. A little ozone in the atmosphere at the earth's surface is stimulating to animals, but a high amount is harmful. There is some experimental evidence that ionization of the air, that is, the conversion of neutral gas molecules into electrically charged ions, may affect the health and the vigour of animals. The height of the ionosphere and ozone layers above the earth is controlled by the intensity of solar radiation, and it is possible that cyclic variations in the height of these layers may affect organisms in ways that are little understood at the present time.

Solar radiation is not received in equal intensities in all parts of the world. Its intensity varies because of inclination of the earth's axis relative to the sun, differences in terrain, amount of cloudiness, and so forth. Solar radiation likewise does not have an identical effect on all species because of differences in their sensitivity to it and because critical periods in their life cycles come at different times of the year. These theories are conjectural at the present time but influence of extramundance forces on animal and plant life should not be dismissed until their potential effects are thoroughly explored.

Population Explosion of Man

The world population of man increased from only 2-3 million at the time of Christ to approximately 1.5 billion in 1960. It then began to explode, reaching 3.6 billion in 1970 and may exceed 6 billion by the year 2000. This irruption is due to reduction in mortality rates without compensating decrease in birth rates.

In his early history, man had to depend on solar energy for the production of both food and heat. With the discovery of life, he used the burning of wood to furnish heat when solar energy was deficient, to cook his food, and to expand his capacity to use machines for locomotion, farming and industry. This allowed his dispersal into parts of the world hitherto closed to him and the production of more extensive food supplies. The next step came with the mining of coal about eight centuries ago and production of petroleum just over one century ago. World consusption of e ergy from fossil has increased about 4 per cent per year during the last century. The increase in world population is doubtless a response in large part to this increase in energy resources, and if atomic energy becomes generally available, there will be no restraint on continued population expansion on this account

Early man through development of superior intelligence and the use of tools soon eliminated predation as a significant mortality factor. With improved medical practice and expertise, many common diseases have been eliminated or at least greatly reduced in incidence or severity. Emigration has extended to nearly all parts of the world, although less extensively, into deserts, tundra, and thick tropical forests. The oceans have no resident human population, although they could conceivably be occupied if there were sufficient demand and engineering ingenuity.

Reproductivity as expressed both in fecundity and success of young in attaining maturity, generally decreases in lower animals as populations expand, but in many species the decrease, especially in fecundity, is insufficient to prevent the populations from reaching thresholds where starvation, disease or competitive struggling becomes disastrous. Such thresholds will be reached also by man unless a determined efforts is mede to reduced fecundity to a point where the population can be stabilized. Man is unique among animals in that his intelligence has evolved to an extent that he can anticipate the future and plan for his welfare, although it remains to be demonstrated whether he has the will and the discipline to do so. It is encouraging that European man's birth rate in 1800 was about 35 per thousand adults per year, while by 1970 it had dropped to about 17 per thousand. This however, is not enough, and annual birth rates in the developing countries remain high, around 45 per thousand.

The average number of offspring per family in 10 European countries averages 2.7; in the United States it has been 3.8. A stable population could be established if the number were reduced to 2.3 per family. However, even when this number is reached, the population will continue to rise for a while. A high proportion of the present-day world population is under 15 years of age. As these young people mature, the size of the child bearing fraction of the population will increase and more than offset the decrease in birth rate for at least one generation.

To check the population explosion, the approach to parenthood is to recognize that it is a privilege rather than a right. If voluntary family planning is inadequate, then stringest government regulation may be required. Continued population growth will lead only to disaster. Involved necessarily with deliberate control over population numbers is control over its genetic make up. Possibilities exist not only to eliminate factors from the germplasm that induce physical and mental defects but also to breed individuals or races with special characteristics that seem desirable. Man has control of his fate in more ways than one.

The population explosion of man is due to elimination of predation, emigration into new habitats, increased survival of young, reduction of disease, increased productivity of food – all related to man's increasing knowledge and technical skill in the exploitation of natural resources. Mankind is headed towards starvation, epidemics and competitive struggling, unless there is determined reduction in fecundity, a reorientation of social and industrial objectives and maintenance of environmental quality. Although man can make more efficient use of the space he has available, the world is still a finite body. Sears (1958) states : Man's physical body occupies space, somewhere between two and four cubic feet. At his present rate of increase it would be less than 700 years, say 22 generations – until there is standing room only, with eachspace of 3 by 2 feet or 6 square feet occupied. On this basis there is a room for exactly 14,646,000 people in each square mile. I have perhaps been over generous in estimating the per capita area but I did wish to leave space enough to permit each individual to reach in his pocket for the rent money when it becomes due. A little after this the hypothetical human population would weigh more than the planet.

6

Geographical Ecology

Introduction

Through the study of geographical ecology we attempt to understand how organisms are distributed over the world and what forces brought about this distribution. The explanation of why a particular species has come to occupy a particular niche in a particular part of the world requires a knowledge of where it originated, how it dispersed, and how it evolved its present adjustments and characteristics. When populations invade new regions and environments and become geographically isolated, they commonly differentiate into new species. Each species thus becomes the product or visible expression of a particular combination of environmental factors, interactions, and locally, and as such the species is the practical taxonomic unit with which the ecologist must deal.

The general laws governing the distribution of animals can be reduced to three simple propositions; a species of animal will be found in any part of the earth having conditions suitable for its maintenance unless (1) its individuals have been unable to reach this region because of barriers; or (2), having reached it, the species has been unable to maintain itself because of inability to adapt to the region or to compete with the other forms; or (3), having arrived and survived, it has subsequently so evolved in the process of adaptation as to have become a species distinct from the original type.

In this lesson we will be concerned both with dispersal dynamics and with the distributional patterns of floral and faunal units and ecological communities that result.

Dispersal Dynamics

Dispersal is the spread or individuals away from their home-sites. Dispersal movements are usually slow, and cover relatively short distances in the life time of an individual. The cumulative result of short dispersals by successive generations, however, may become conspicuous in the course of years, decades, or centuries, when it constitutes range expansion of the species into a new habitat or area.

Under entirely natural conditions he rates of dispersal of all species, once a barrier is passed, would be similar, varying only with respective reproductive potentials, characteristic speeds of locomotion, and relative abilities to find unoccupied niches, overcome competition, acclimatize to new climatic conditions, and acquire new behaviour patterns.

Manner and Means of Dispersal

A common method of achieving dispersal is the broadcasting of enormous numbers of eggs, spores, encysted states, or young so that they scatter into a wide variety of places in a more or less random manner. Those that come by chance into suitable environments persist and become established; those that enter into unfavourable locations are destroyed or never develop. Broadcasting is a wasteful procedure; it is especially characteristic of aquatic species. The fresh water clam annually produces hundreds, perhaps thousands, of eggs. Only two fertilized eggs need mature that the two parents be numerically replaced when they die, and thus the population of the species be maintained at a constant level. Constrastingly, in those forms, such as birds and mammals that have developed a high degree of parental care, the number of eggs or young annually produced is commonly a half-dozen or less, and the off-spring exercise considerable discrimination in their choice of suitable habitat.

Although the dispersal of broadcast offspring is not under the control of the parents or the young, it is not often truly random. Water and wind currents and other agents of dispersal many channel eggs and spores in restricted directions. Such dispersal is described as passive conveyance. In streams, all agents of passive conveyance, except some other animals, direct dispersal downstream. Upstream dispersal must be the result of active locomotion.

Eggs of insects, snails, fish, or other aquatic organisms will sink unless they are buoyed up by currents, possess flotation mechanisms, or are attached to some floating object. Logs, masses of vegetation, and other debris are sometimes torn loose from the banks of rivers and float out to sea carrying the smaller animals attached to or trapped on them. It is estimated that may become attached to mammals, resident birds, or insects for more local dispersal. A species of Succinea snail, native only to St. Croix and Puerto Rico, was found alive in the feathers of a bobolink shot in Cuba. Ferrisia snails have been found attached to the wing covers of aquatic beetles and sphaeriid clams clamped on their legs. These insects occassionally fly from one body of water to another Crustacean eggs have also been found viable after passing through the digestive tract of different kinds of birds. Plant Seeds may be dispersed thousands of kilometers in the digestive tracts of birds. Fruits become conspicuous, abundant, and nutritious, thus attracting birds at the proper time to aid in their seed dispersal.

Some dispersal movements are determined by the manner in which the animals respond to environmental factors. Such directed movements, or taxes, may induce the dispersal of animals upstream positive to current rather than downstream.

Trial and error is involved to a considerable extent in the dispersal movements of animals. If, initially, an individual progress into an unfavourable habitat, it may be able to withdraw and proceed in a new direction. This may continue many times until by chance it discovers an area that is favourable. Trial and error movements are manifested in all groups of animals. In higher types, and even in many lower types, the individual may learn by experience and reduce the number of false trials that it makes. Thus initially random movements may eventually evolve into directed behaviour patterns,

Barriers

Dispersal continues until a barrier that makes further movement difficult or prevents successful colonization is encountered. Barriers are of different sorts and are classified as physiographic, climatic, and biotic. The difficulty of surmounting or bypassing any type of barrier varies greatly among species. River valleys, for instance, may be barriers to animals frequenting mountains ; but to lowland species, river valleys are important dispersal highways.

To fresh-water organisms in river and lake habitats, intervening land masses are usually effective barriers. The headwaters of different river systems may lie only a few kilometers apart, yet contain quite different species because the continuous water route down to the mouth of one river and back up to the other may be a distanae of several hundreds or thousands of kilometers. A waterfall may be a barrier for all non-flying aquatic species, and even riffles or swift water may prevent upstream dispersal of pond or lake species. The salt water intervening adjacent rivers flowing into the sea is a barrier to most fresh-water forms that might otherwise invade one river from the other.

Terrestrial organisms are hemmed in by a great variety of barriers. The oceans constitute the major physiographic barriers since they separate the faunas of the several continents and isolate islands from each other. Lakes are not effective barriers because they can be readily skirted, but wide rivers coursing long distances between banks of dense vegetation as does the Amazon river through the tropical rain forests of South America, may limit the range of forest mammals, butterflies, flightless beetles, land snails, and even birds. The Grand Canyon of the Colorado River separates the ranges of the Kaibab and Abert squirrels and several other species, even in country that is semi-arid and open. Mountains are sometimes considered barriers to lowland species. This is only true if the change in climate and vegetation that such barriers produce are unfavourable to the species.

Deserts are important climatic barrier since they are hot and dry. Temperature affects animals directly, since species have definite limits of tolerance, comfort, and efficiency. Precipitation is important because it controls the type of vegetation that occurs in a region, and the animals adapted to that of vegetation. A low relative humidity may directly limit the dispersal of moist-skinned species. Excessive solar radiation limits some species to forest habitats, excluding them from open country; without recourse to shade, such animals would experience overheating and critical loss of water from the body. Short photoperiods may limit distribution north-ward during the winter, especially if low temperature obtains, and may influence southward distribution of normal animals in the summer. The length of the season between spring and autumn killing frosts, or between dates when its limits of temperature tolerance are reached in the spring and autumn, may determine whether or not a species can complete its life cycle at a given latitude. A corollary consideration is whether the accumulation of heat is sufficient to furnish cold-blooded animals and plants sufficient energy for growth and reproduction.

Biotic barriers consist of changes in vegetation, food, competitors, and predators. The adaptations and behaviour patterns of many animals fit them to niches in specific types of vegetation; should the previously amenable vegetation change, the animals may have great difficulty in adapting to it. Tree squirrels, for instance, are replaced by ground squirrels in prairies and deserts. Most animals have a great adaptibility to food, so food is not so limiting a factor. Some insects, however, such as certain aphids, are narrowly limited to particular species of plants as a source of food. Where their food plant is not present, they cannot exist.

Function of competition between the two species Either species can live in the range of the other, but in the North, the house were usually wins in competition for territory and nest-sites. In the South, the Bewick's wren is more successful. Predators, such as the great horned owl and the swifter hawks, tend to confine the bob-white to forest-edge habitat, where it is less vulnerable to attack. Trypanosome parasites carried by the tsetse fly are effective barriers against successful introduction of domestic ungulates into certain parts of Africa, and the rabbits introduced into Australia have limited the range and greatly reduced the abundance of several species of native marsupial.

Dispersal of Young

The dispersal of a species is primarily accomplished in the immature stages. This is obviously true of eggs and spores, but banding and marking studies have shown that among the higher animals – birds and mammals – it is also the young which dispersed the species. Once a bird has reached sexual maturity and nested. It has strong tendencies to return to the same area in following years. The distribution of young birds is not random, however, as they tend to return to the general vicinity of their birthplaces rather than uniformly over the range of the species. Thus only 0.5 per cent of 557 adult house wrens recovered a year after banding nested farther than 33 km (2 miles) from the site where they had nested the year of banding, but 15 per cent of the 182 birds banded as nestlings were recovered at greater distances, the longest of which were 32 km (20 miles) 56 km (35 miles), 80 km (50 miles) and 1120 km (700 miles). Dispersal distances for young of other species are proportionally comparable. Apparently the young are at a disadvantage in the competition for nesting territories with adult that have previously nested in the area, hence are the ones forced to go elsewhere.

Of small mammals, it is characteristic that once an individual has selected a homesite, it rarely leaves it for another. It was observed that in the months following the time at which they had been captured, marked, and released, 95 per cent of 133 adult woodland white-footed mice resumed habitation within 183 m (200 yd) of the site of capture.

Rate of Dispersal

If dispersal from birthplace were typically limited to one direction, then a simple mean of the distances to which the young disperse before they breed would give the dispersal rate per generation. It is the case, however, that dispersal proceeds peripherally in all available directions and may extend to surprising distances.

The area exposed to invasion and the average time required to saturate that area increase proportionally as the square of the linear distance (d^2) from the centre, since the total area within which the individual could settle is πd^2. A number or equations have been suggested for computing mean dispersal distance. Consider the data on house wren nestling recovery distances, presented above. Excluding the truly extraordinary distance of 1120 km, and observing that only about 93 per cent of young female wrens nest when they are 1 year of age, we compute an annual dispersal rate of approximately 8 km (5 miles) for this species. The mean dispersal distance of one group of 154 young woodland white-footed mice (Burt 1940) is about 176 m (192 yd). However, mice bron in the spring mature sexually very quickly and breed in late summer or autumn, so the annual dispersal rate must be somewhat greater than this figure indicates. The annual dispersal of marked snails resulting from random movements along a linear bank of uniform habitat averaged only 5.5 m (6 yd) (Goodhart 1962).

Our dispersal rate data so far have described the outward diffusion of a local population through an area already occupied by the species. Once it has surmounted a barrier the dispersal rate of a species into a suitable area previously unoccupied by it should be faster. The European starling was introduced into North America about 1890. From a central locus around New York City it spread at an accelerating pace until, in 1940, it had become established over 6,500,000 km² (2,500,000 mi²), a mean rate of about 130,000 km² (50,000 mi²) per year. With amelioration of the climate in Finland during recent years, the lapwing spread northward between 1899 and 1954 at a mead annual rate of 7 km (4.3 miles); the roe deer, between 1850 1945, at a mean annual rate of about 95 km (58 miles). The Norway rat invaded southwestern Georgia and virtually displaced the previously established black rat at a rate of about 430 km² (167 mi²) per year (Eckh 1954). By way of contrast, it has taken the fresh-water amphipod Gammarus pulex the last 6000 years to disperse across 12 river systems from southern England into Scotland.

Causes of Dispersal

The reproductive rate of any species is so great that if all offspring survived the world would be overrun with that species within relatively few generations. Because species produce a surplus of young in most years, there is continuous pressure on individuals to move into all suitable niches, and to seek out new areas in which to settle. The impact of large numbers of individuals struggling for survival is described as population pressure, and is doubtless the most potent force inducing dispersal. It should be recognized, however, that population pressure is not uniformly constant year after year. When because of poor breeding conditions or catastrophe there is a reduction in the over-all population of a species, that decimated species may withdraw into its optimum habitat and be less put upon to exploit new or less desirable areas. In years favourable to the production of large surpluses of young, a species will often be found in less favorable habitats, even regions it would not otherwise occupy at all. The broadcasting of eggs or offspring, or the passive conveyance of the them to other regions, varies directly with the size of population producing them, and is hence as much an expression of population pressure as the active search for new areas engaged in by individuals under their own locomotion. In addition to relieving population pressure, dispersal is advantageous to the species for faster spread of mutations through the entire population and for occupation of new or depopulated areas. The tendency of disperse may even have been selected for in evolution so that it has become innate in young animals.

Animals cannot disperse successfully, if at all, into new areas to the characteristics of which they are not structurally, functionally, and behaviourally adapted. If an areas, which has excluded a species, changes so that the species is then adaptive to it, that species can successfully invade. The American robin, song sparrow, chestnut-sided warbler, house wren, and prarie horned lark have invaded Georgia only in recent years as the logging of forests, initation of early seral, grassy, and shurbby stages, and extensive general cultivation of the land have produced habitats meeting the requirements of these birds. Likewise, if an area remains unchanged but the species acquires new structural, functional, or behavioural traits by which it can adapt to that area, the species can invade the area. If the food supply fails, homesites or vegetation are destroyed, or a pernicious change in climate occurs there animals may be forced to leave an area to which they were well adapted and disperse, more or less temporarily, into an area to which they are less well adapted. The snowy owl, for instance, depends heavily on lemmings and mice for food in its usual range, the Arctic tundra. In correlation with the cyclic decline of the lemming population there, the owls invade our northern states.

Dispersal Pathways

An understanding of how animal and plant groups have dispersed over the face of the earth involves the controversy as to whether all the continental land masses were at one time intimately connected and if so, how long ago they drifted apart to their present locations. Separated continents need at least temporary bridges if there is to be dispersal and interchange of fauna between them.

Continental Drift Theory

This theory postulates that throughout the Paleozoic and much of the Mesozoic the presently distributed continents were grouped into two great land masses : a northern Laurasia was separated from a southern Gondowana by the vast sea of Tethys, although narrow connections between the two might have occurred for short periods. In the Jurassic of Lower Cretaceous the land masses fragmented, and the fragments subsequently drifted apart Laurasia is supposed to have split into North America, Greenland, Europe, and most of Asia; Gondwana, into South America, Africa, Arabia, Madagascar, India, Australia, and Antractica. This theory is suggested by the shapes of the continents that could conceivably be fitted together; by the characteristics of the Atlantic Ocean basin that make it appear to have been formed by a rifting apart of land masses; by new paleomagnetic studies; by the similar geological stratigraphy shown by characteristic assemblages of rocks and similarity in their ages, and by invertebrate fossils in Africa and South America; by some similarities of present-day fauna and flora at the same latitudes on different continents; and by the difficulty tropical species have of dispersing over arctic land bridges.

Evidence in favour of the continental drift theory has been increasing in recent years in fact, drift is apparently still going on. Deep sea drilling shows that sea floor has been spreding outward from the mid-Atlantic ridge at a uniform rate of 2 cm/yr. However, the continents have certainly been separated thoughout most of Tertiary time and perhaps also the latter part of the Mesozoic so that dispersal of modern organisms over he world must have taken place in other ways.

Land Bridges

An alternative theory is that the continents have been situated in their present positions throughout geological time. They have changed their boundaries with rise and fall of ocean levels and with upheaval and subsidence of land masses, but there has been no significant lateral drifting. This theory was generally held to be correct by most geologists and biogeographers up untill the last few years. Regardless of the ancient history, the continents have been largely separated during the last 50 to 100 million years, and movements of animal and plant groups during this latter period have been most important in determining their present distribution.

The separated continents, however, have been connected by at least temporary land bridges at various times during the period. A land bridge now connects North and South America at the isthmus of Panama, and Eurasia and Africa are connected by the isthmus of Suez. One of the best known and generally accepted land bridges of the past, but not now in existence, connected. Asia and North America cross the Bering Strait and Bering Sea. Possibly bridges or extensive archipelagos occurred in tne Atlantic Ocean area in the past that permitted dispersal of organisms not possible at the present time, but concepts of this sort are highly hypothetical at the present time.

Land bridges may exist for only a short time in the geological sense, but they serve as important dispersal routes for those land animals and plants able to cross them. When a land bridge allows free passage of most animals and plants in either or both directions, it is called a corridor. If the land bridge is narrow, has an unfavourable climate, a lack of suitable niches, or too many competitors, it is called a filter—only a few specially adapted species are able to pass over it. A sweepstakes route of dispersal is one over which only a few species pass, more or less by chance. Because of its general unfavorableness, it is generally only one-way. This island-hopping pan-tropical dispersal of organisms from southeastern Asia through the south Pacific has been accomplished by relatively few species; the more distant the island, the fewer the species that have reached it.

A glance at a globe quickly show that the continental land masses are concentrated mostly in the northern hemisphere. There are three broad land extensions southward below the equator : the Malay Peninsula, East Indies, and Australia; Africa; and South America. The southern hemisphere otherwise consists largely of vast expanses of oceans. On a land route, North America intervenes South America and Eurasia. Eurasia is the largest continent, is central to all the others, and always has had a great diversity of climate and terrain. The size, arrangement, and positions of the continents are of importance to interpretation of the past evolution and dispersal of animals.

Considerable evidence indicates that some large groups of animals, notably mammals, first evolved in Eurasia and then spread to other parts of the world. North America was a less important centre of origin and dispersal. Warm, moist, uniform climates prevailed in the early Carboniferous, Jurassic, mid-Cretaceous and Eocene. During periods of continental emergence, climates on the northern continents changed from moist and warm to more arid and cold. Periods of aridity and glaciation are known to have occurred in the northern continents during the Permian, at the end of the Tirassic, at the beginning and end of the Cretaceous, and during the Pleistocene. Animals adaptively limited to moist, warm environments were restricted to tropical regions or dispersed outward into the southern land extensions. New more advanced animal types adapted to new conditions in the north appeared. Monotremes entered Australia at an early date. Marsupials probably originated in Eurasia and dispersed into Australia and South America during the late Mesozoic or early Tertiary, although there is no fossil evidence for these supposition. An oppossum still occurs in North America, but no marsupial is now present in Eurasia. Marsupials did not get into Africa, but are present in South America and have adaptively radiated into a variety of forms is Australia. The predominance of marsupials in Australia is probably due to their coming over at an early date on a land bridge on their chance success in surmounting the sweepstakes route bridge was an important route of dispersal between Asia and North America considerably more study is required before we will satisfactorily understand how this was accomplished by various kinds of animals.

Centers of Origin

The tracing of dispersal routes presupposes a starting point where the taxonomic group, whatever its size, first evolved. These starting points are called centers or origin. Various criteria for determining centers of origin have been suggested, but caution must be exercised in applying them. Of the many criteria proposed, the following two are especially important, although neither one is infalliable.

1. Location of greatest differentiation of the type or the greatest variety of endemic races, species, and general including primitive forms or fossils. The older a group is and the longer it has occurred at a particular location, the more chance it has had to radiate into different habitats, become isolated, and evolve into new varieties. However, a shift of climate or a drastic change in physiography may render an original locality unhabitable and the group moves elsewhere. Also, a group losing its vitality may contract its range into some area other than the one in which it originated.

2. Continuity and convergence of lines of dispersal. Dispersal from a centre radiates in all directions in which conditions are favourable and until insur-mountable barriers are encountered. Lines of dispersal may be readily distinguished where one or more taxonomic characters can be traced from primitive or generalized types through more and more specialized types the greater the distance involved. However, once a species filters through a break in a barrier and invades an extensive, new, and favourable habitat, there may be increased evolution of new types, and a secondary dispersal centre formed. This has happened repeatedly in the geological past, so there is often difficulty in distinguishing which centre is the original one for a group.

Island Dispersal

Dispersal of animals from continents to islands and from one island to another presents special problems. Many of the world's prominent islands occur on the continental self and are separated from the mainland only by shallow seas. At times of land emergence, as when glaciers look up quantities of water as snow and ice enough to lower the level of the seas, these islands becomes connected by land bridges to the mainland, and dispersal of forms occurs. The British Isles have been thus connected to Europe; Japan to Korea and Siberia; Sumatra, Java, and Borneo to Malaya; New Guinea and Tasmania to Australia; and Newfoundland to Labrador. On the other hand, the Bermudas, Azores, Hawaiian and other small Pacific island, and possibly New Zealand, the West Indies, and Madagascar, could not have mainland connections and thus must have received their present faunas by some means other than overland dispersal.

Islands adjacent to continents, unless very small or long separated, generally have faunas similar to that on the nearby mainland. Oceanic island are more difficult to colonize, however, and often have unique unbalanced faunas or chance assemblages of species. Mammals, amphibians,and fresh-water forms are often absent or scarce. Flying birds, bats and insects blown in by strong winds, and small invertebrate forms readily transported on rafts are better represented.

Since islands are colonized with difficulty, the number of species on them is generally small compared with the mainland, and interspecific competition is reduced. A single established genus or family on islands free of close competitions may adaptively radiate into new riches to form a variety of species or races. Animals on island are often smaller in size than their relatives on the mainland, for instance ponies on the Shetland Islands, Key deer on the Florida Keys, and the gray fox on Catalina Island. Islands are generally too small to support adequate populations of large animals requiring extenive home ranges. At the same time, there may be selective advantages operating in evolution for becoming smaller. Because of lack of competitors and predators, primitive animals isolated on islands may survive long after their relatives on the mainland have perished, as has, for instance, the reptilian Sphenodon on New Zealand. Confinement of a species to a limited range permits extensive inbreeding so that the population becomes more homozygous in its various genetic traits, hence much less adaptable to new situations than are larger heterozygous populations Traits that would be eliminated by predation pressure on mainland sometimes becomes established in populations on islands. All these conditions render oceanic island species liable to extinction by invasion of mainland forms, and renders them impotent to reinvade the mainland.

There is continual invasion of individuals of the same or new species into island; at the sametime there is continual extinction of individuals and species. A biotic equilibrium is reached only when the two rates are equal. The relation between rates of invasion and extinction varies with the distance from the mainland. Successful colonization after invasion is more likely the higher the ratio of birth rate to mortality rate The number of established species commonly varies.

A continuous land connection occured between North and South American in late Cretaceous time. Ancient types of mammals, birds reptiles, amphibians, insects and other groups got into South America over this land bridge from the north and differentiated into distinct families and other toxonomic categoeis. During most of the Tertiary, no land bridge existed between the two continents; although there were scattered islands separated by relatively narrow water gaps which some groups, particularly birds, may have been able to use. The land connection now in existence was apparently formed in the late Pliocene or early Pleistocene.

Previous to late Pliocene there were about 29 families of land mammals confined to South America and 27 confined to North America; the two continents had not more than one or two families in common. During the Pleistocene, after the land bridge had been in existence for some time, 22 families were represented on both continents; 7 had dispersed from South to North America, 14 from North to South America, one is of uncertain origin. Some families have become extinct. The present faunas of the two continents contain 14 families common, 15 families found only in South America, and 9 families found in North America. Thus there has been considerable dispersal over the land bridge in both directions, but not a complete exchange or unification of faunas (Simpson 1940).

The modern fauna of North America is thus derived principally from (1) Eurasia, (2) South America, and (3) by autochthonous development. By autochthonous we refer to species evolved from very old indigenous types that may or may not be represented by related forms on other continents. The proportion of any local fauna that is derived from one or the other of these sources varies with each geographic area and with each group of animals. Northward on the continent and in the western mountains, the Eurasian bird element is strongly represented. The South American element becomes greater southward, especially in the lowlands of California, Mexico and Central America. Many Eurasian forms have dispersed through North America into South America, but no modern forms, at least, of South American origin have been able to invade Asia through North America.

Migration

Migration, like dispersal, involves movements and the invasion of new areas. Migration, as here defined, differs from dispersal in that it is a periodic movement back and fourth between two areas. In contrast, dispersal is a one-way outside movement. Migratory invasions of areas are temporary and repetitive, but invasions resulting from dispersal may be permanent.

Migration is best known in birds, as an invasion of breeding area alternating with an invasion of wintering area, annually. Representatives of other groups of animals also migrate, particularly mammals, fish, and insects. Migration may be classified as annual, diurnal, or metamorphic. Annual and diurnal cycles are correlated with the two most pronounced time cycles in the physical environment. Metamorphic migrations are movements from one habitat to another in different stages of an animal's life cycles.

Annual Migrations

Annual migrations may involve a change of latitude, or altitude, or be more local in extent.

Latitudinal migrations may traverse only a few kilometers or may extend almost from pole to pole. In terms of their occurrance in an area bird species are described as permanent residents species represented in an area throughout the year even though, some individuals migrate; summer residents, species present only during the warmer part of the year, which includes a breeding season that may extend from early spring to late autumn; winter visitors, species present only during the winter or non-breeding period; transients, species present only during migration periods, neither breeding nor wintering in the area; and accidentals, species that are rare or irregular in occurrence. The bird population in most communities reaches peaks during the vernal and autumnal aspects, as transients arrive and temporarily swell populations. The autumnal peak usually the greater because adult populations are incremented by the larger number of young birds hatched during the breedings season Latitudinal bird migrations also occur in the southern hemisphere but are less conspicuous because of the small continental land masses there and small populations of birds involved.

Altitudinal migrations are movements of no more than a few kilometers up and down the slopes of mountains.

By descending to lower altitudes in the autumn, an organism obtains some of the same benefits secured by those species that undertake latitudinal migrations - less snow, higher temperatures, and more food. Birds restricted to alpine habitats in the summer are common winter residents of lowland areas. Some of the larger mammals, such as the mule deer and the American wapiti have very regular migration habits in respect to herding, timing of movements, and migration routes. They move to the high alpine meadows soon after the vegetation renews its growth in spring, and come back down to the valleys in the time avoid the deep winter snows of the higher slopes.

Local migrations do not necessarily involve a change of latitude or altitude and are often quite limited in distance covered. However, in tropical grasslands and savanas, where wet and dry seasons bring great changes in available water, vegetation, and food, there is a great exodus of both mammals and birds during the dry season and an influx during the rainy season.

The Atlantic salmon, after reaching sexual maturity, may ascend fresh-water streams in subsequent years to spawn and return each time to the sea. Many deep water fishes spawn annually in shallow waters and then return to deep water again. Turtles come onto the land to lay their eggs, snakes disperse from their winter dens with the advent of warm weather in the spring; tree frogs go to small pools to mate and spawn; and resident birds move onto their breeding territories.

Insects perform regular migrations both into hibernation and out of hibernation. In the autumn, forest species migrate downward and may be found in peak numbers first in the shrubs, then in the herbs, then in their hibernacula in the ground. Many insects of the forest-edge, meadows and agricultural crops also hibernate in the forest, usually a few meters in from the south-exposed edge, where they derive some heat from the winter sun. These insects usually migrate into the forest in the same stratum, herb or shrub, in which they occur during the summer, then downward into the soil. These flights into the forest occur with declining temperatures and sometimes include spectacularly large numbers of individuals. As they come out of hibernation in the spring the direction of movement the insects take is just the reverse that taken in the autumn, that is, upward into their proper stratum, then horizontally outward into open country.

Daily Migrations

Ascent of plankton toward the surface at night and descent to greater depths during the day occur both in the sea and in lakes. The lake dwelling culicid larva chaoborus lies on the bottom during the day, but becomes pelagic at night. Snails, slugs, and millipedes in the deciduous forest lie quiescent under logs or litter during the day, but come out at night to crawl around on the forest floor or even climb up on the vegetation to a height of perhaps a few meters. Although these excursions are restricted in range, they are more or less regular and periodic, and may be thought of as migrations.

Metamorphic Migrations

Acquatic larvae and naiads of several orders of insects eventually change into adults that leave the male aquatic habitat and become aerial. The adult stage is often short-lived. The eggs are deposited in water, or the immature stages return again to water to begin the cycle over again. The length of the cycle may be part or the whole of a year, or a longer period. The seventeen-year cicada is a well-known insect whose nymhps spend 17 years underground feeding on juices from the roots of trees. The adults appear above ground in large numbers in late May, mate lay their eggs on twigs of various trees, and then disappear, all in a few weeks. The eggs hatch in about 6 weeks, and the nymphs drop to the ground and bury themselves for another long period of years.

The anadromous Pacific salmon ascends freshwater streams but once to spawn. The breeders die; it is their nff-spring that return to the ocean to develop for a period of years before they make the migration. Many adults return to spawn in the same stream or even in the same headwaters where they were hatched. The Pacific salmon apparently uses the sun as a compass mechanism to find its way from the ocean to the stream, then swims upstream against the current, rejecting all odors until it arrives at the home tributory to which it was initially conditioned. The two species of catadromous eels, of the western hemisphere and of Europe, migrate to the open sea in the region northeast of the West Indies in order to reproduce. The immature stage of the eel not the adults, returns to the two continents.

Ecesis

Dispersal or migration of individuals into new areas is without great ecological significance unless those individuals become established and can build up significant populations. The process by which organisms become established in new areas is called ecesis. Ecesis will occur if a species disperses into a habitat favorable to it and can secure its proper niche or become adjutsed to a new niche, new competitors, predators, parasites, and disease organisms Ecesis, to go to completion, requires first the establishment of individuals in an area, then the growth of the population that they form, and finally or simultaneously, the maturing of community structure with the invasion of many other species.

Temporary ecesis is the rule with migrant species, as the periodic change of location is normal in their life behaviour. Ecesis as range expansion is not surely permanent until the species demonstrates that it can survive critical years in weather cycles. For instance, northward dispersal of tropical forests is thwarted by frosts that occur only at rare intervals. Insects may continuously expand their ranges for a period of years, only to be forced back hundreds of kilometers by a severe drought or cold spell.

The total community is an aggregation of many species. With seasonal onset of favorable weather or when a new habitat becomes available with progressive development of substratum, microclimate, and vegetation, the coordinated invasion and ecesis of these many species also follows the sigmoid or logistic curve. When, however, the new habitat is fully developed and ready for colonization by a group of species, as for instance islands the lower self-accelerating portion of the sigmoid curve may be eliminated and only the upper of self-inhibiting portion expressed Furthermore, colonization curves for insects arriving on small mangrove islands in the Florida Keys, after removal of the original fauna by fumigation, show that a larger number of species invade than can become established Extinction of less aggressive or more poorly adapted species results from various causes, including competition and predation. The asymptote of the curve is reached when the species present establish interactions between them so as to form a closed or stabilized community that discourages further invasion. Thus the process of growth or ecesis is much the same whether it is at the level of the individval cell, organism, species population, or community.

With succession, a series of communities become established one after another. This is best represented by a sequence of sigmoid curves, each indicating a particular stage in the sere. As a new community invades, the previously established one disappears, probably at a rate indicated by a reverse sigmoid curve. Ordinarily a larger number of distinct plant communities can be recognized in a particular sere than animal communities. The ecesis of a single animal community may overlap several plant stages. Thus the deciduous forest biociation begins its development on the Mississippi River floodplain with the establishment of the first trees in the cottonwood-willow associes, continues to develop through the sugarberry-sweet gum, oak-hickory associes, and into the climax oak-tuliptree association. However as the climax reaches its full dominance and all remnants of earlier plant stages are eliminated, the decrease in diversity of plants is often correlate with a decrease in species diversity of animal. The ecesis of distinct animal communities does not correspond, therefore, with ecesis of distinct plant communities, but the ecesis of animal communities is correlated only with the development of distinct types of vegetation. A background of understanding of the processes of dispersal and ecesis and the problems encountered by species in their invasion and establishment in different parts of the world is desirable for proper evaluation of present-day distribution of organisms and the distributional units that are recognized.

Distributional Units

Only a few groups of animals are cosmopolitan or nearly so, in having a worldwide distribution, notably cyprinid fish, frogs of the genus, Rana, colubrid snakes, passerine birds, rodents, and man. Most organisms are restricted in distribution because of barriers of one sort or another or because of past history of origin and dispersal. An understanding of their present-day distribution rakes us into biogeography, climatology, and paleoecology. Major units of distribution are the zoological regions, areas defined largely by the past and present relations of the continents to each other. Each region may then be subdivided into faunal or ecological units, depending on the criteria used.

Zoological Regions

Zoological regions have long been separated one from another by water, mountain, or desert barriers so that they have evolved distinctive and characteristic orders and families of animals.

The Australo-Popuan region has been isolated the longest and is marked by the absence of most groups of placental mammals, although some have been introduced. The original mammalian fauna, derived from Eurasia when Australia and New Guinea (Papua) connected to that continent by a land bridge, consisted of monotremes and marsupials. Murid rats and mice and bats invaded later and in a different manner. It is of interest that woodpeckers among birds, true frogs among amphibians, and fresh-water fish are absent or scarce.

The Neotrophical region contains the largest number of unique endemic forms of birds and mammals perhaps because it is the farthest by a land route from the oriental tropics where the major taxonomic groups probably originated and because the region is almost completely surrounded by water. The occurence of many related forms in South America, Africa, and Australia, and the absence or poor representation of these forms. Even as fossils, in North America and Eurasia may be explained either by the continental drift theory or by connections in the geological past with a much warmer Antarctica continent. For instance, the ratite bird, rhea, and the lungfish, Lepidosiren, occur only in South America; the ostrich and the lungfish Protopterus only in Africa; and the emu and cassowary birds and the lungfish Epiceratodus only in Australia and adjacent islands.

The Ethiopean region is noted especially for giraffes, antelopes, zebras, elephants, and other ungulates that wander around in large herds; by the large hippopotamus and rhinoceros; by a number of cats and dogs; and by the primate lemurs, baboons, macaques, chimpanzees, and gorillas. There are several endemic families of birds and other forms. Absent are deer and bears among mammals and salamanders and tree frogs among amphibians.

The fauna of the Oriental regions shows relationship to that of the Ethiopean region in several respects. India was connected with Africa at various times.

Faunal Groups or Elements

In all the systems above described, faunal distribution has been analyzed in terms of geographic areas, and the chief criterion for determining the limits of an area has been that of distinctiveness between the fauna of different areas. The fauna or flora is analyzed in terms of groups of species with similar centers of origin, dispersal routes, geological histories, and habitat preferences. The objectives are to learn the place where and time at which these groups of species originated, how the groups came to occupy a particular part of the world, became adapted to live in new environments, and how they evolved into the present-day living species. A type of vegetation, life zone, or biotic province may contain species from several different faunal elements that have come to live together.

Biomes

A still different approach for the study of distribution is the ecological one, involving the recognition and classification of natural communities. A number of different ecological systems have been suggested but we will here be concerned only with the biome concept. The systems described earlier in this section are zoogeographical, in their attention to centers of origin, dispersal routes, and evolution. The biome system emphasizes community dynamics and environmental relations. This does not mean that zoogeographers do not consider the relations of climate and terrain in controlling dispersal and evolution, nor that ecologists are not concerned with the geological history of the forms with which they deal, but the viewpoints and objective are different.

Basic to the understanding of the biome system is the recognition of communities composed of characteristic combinations of animal and plant species, of successional relations between communities and that succession in all local habitats eventually converge into a climax community pattern, the most important ecological characteristic of geographic area.

A biome is a biotic community, characterized by distinctiveness in life-forms of the important climax species. On land, the most important climax species are usually plant dominants that occur in distinctive vegetation and landscape types; in the ocean, the important organisms that define biomes are usually the predominant animals which are sometimes also dominants.

Seral communities are developmental stages and are consequently considered part of the biome. The animal and plant constituents of seral stages are, however, more widely distributed than are species belonging to climax communities, since the habitats in which they belong are more nearly alike in different parts of the world than are the habitats that contain the climax. Seral species are not generally useful, therefore, to defining the limits of biomes. The majority of animal constituents of the climax community however, are characteristic only of the climax vegetation or habitat and therefore of restricted distribution.

The principal biomes of the world, or biome-types of some investigators are the following :

Terrestrial	**Marine**
Temperate deciduous forest	Oceanic plankton and nekton
Coniferous forest	Balanoid-gastro-pod-thallophyte
Woodland	
Temperate shrubland	Pelecypod-annelid
Tundra	Coral reef
Grassland	
Desert	
Tropical savenna	
Tropical forest	

7

Evolutionary Ecology

Background

We have been mostly concerned with the factors as to how organisms are adapted morphologically, physiologically and behaviourally to their environment. We are now going to discuss how these adaptations have evolved because of their survival value to the species. It is the consideration of these factors that constitute evolutionary ecology.

Evolutionary ecology is concerned not only with speciation and its accompanying niche segregation but also with the manner in which all ecological structure and function has evolved. Part of this evolution is biological in that species characteristics are transmitted genetically by means of chromosomes and genes, and part of it is cultural in that behaviour and social systems or ethics are transmitted through the training and learning experiences of the young. Human culture, including ethics and religion, has its a evolutionary roots in the behaviour and social systems of lower animals. Natural selection affects behaviour and social patterns as it does structure and function, to render the organisms more fit and able to survive in specific environments and to live together in particular relations with one another. In this chapter we will consider ecological niches and taxonomic species, how they have evolved together to give structure to the community, and how behavioural interrelations between individuals have evolved social patterns, ethics and culture, culminating with man.

The ecological niche is a particular combination of physical factors (microhabitant) and biotic relations (role) required by a species for its life activities and continued existence in a community.

Hutchinson (1957) suggested treating the environmental variables, both physical and biotic, affecting a species population as a set of coordinates. The ranges of these variables within which the species can survive and reproduce then define as a dimensional abstract volume or hypervolume, which he terms the fundamental niche, If two species in a biotope overlap in their fundamental niches, only one species may survive in stable interaion in the volume overlap where they are in direct competition. If the two fundamental niches differ in part, the two species can survive together and coexist in the biotope. The realized niche of each species is, therefore, restricted to that portion of its fundamental niche to which it is better adapted. A community can thus be received as an assemblage of species in a biotope that have divided up the 'niche space' (the full range of environmental variables affecting those species in the community) so that they occupy niche volumes that differ, though they may overlap.

The restriction of a species to particular niche depends on its structural adaptations, physiological adjustments and tolerances, and behaviour patterns. At the same time, living in a particular niche brings increasing evolutionary perfections for living there rather than somewhere else. The variety of species living together in different niches gives ecosystem species diversity. The concepts of species and niche can hardly be considered one without the other, and both concepts need to be approached from the point of view of evolution and adaptation.

Behaviour Adjustments

Somehow an animal must get into a favourable niche out of the multitude of niches available to it. Doubtless, some animals accomplish this passively by the random dispersal of spores, eggs or larval stages, some of which by chance reach favourable locations, there to mature and survive. When immature stages of two or more species find a particular location favourable, that species which settles there first, in the greatest numbers, or grows faster is likely to acquire dominance over the area and force the other species to take a marginal position of disappear.

Higher animals with their more complex nervous systems, greater intelligence, and sense organs more highly evolved, are equipped to search actively for and recognize niches suitable to them either by sight, smell, contact or other means. For instance the intricate migratory behaviour of birds and other animals is such that they seek out nearly constant climatic environments throughout the year. Migratory behaviour may have evolved as the result of groups being unable to solve the problem of niche isolation locally.

The apparent ease and speed with which a new generation of individuals discriminates a hospitable niche means that they must in some way recognize the merits of that niche by definite characteristics of it that may be in the nature of sign stimuli Such characteristics are usually prominent, though they need not necessarily be the most essential features of the niche.

Probably most animals exercise a deliberate, although not necessarily conscious, evaluation process in choosing one niche from those available. This has been tested experimentally by exposing the animals to gradients of environmental factors, either in the laboratory or field. A variety of apparatus and procedures is available for such tests. Usually there is a coincidence of the species experimentally ascertained preferendum and its natural preferendum.

Where two species with similar niche requirements come into competition, one species must possess better adaptation to it if it gains full possession of the niche to the exclusion of the other species. Preadaptations or possession of suitable adaptations also appears a necessary prerequisite for a species to invade a new niche or habitat and successfully displace another species already occupying it, as frequently occurs. The search for such microadaptations as might give one species an advantage over another in competition for a particular niche, is a real challenge to ecological research.

Inheritance

The fact that all individuals of a species behave almost identically in many of their activities indicates that these behaviour patterns or instincts are in some form passed on or inherited from one generation to the next. Behaviour that is learned after birth is much more variable between individuals. Behaviour patterns are rooted in the structural arrangements of neurons and synapses and in the functions of hormones and enzymes. Once a stimulus is received a definite action results. Predisposition for a species behaviour patterns could well be inherited genetically through chromosomes and genes like any other structural or functional characteristics. These inherited behaviour patterns are doubtless subject to evolutionary development as much as are structural and functional adaptations; indeed, the one may have developed synchronously with the others.

There may be transmission of behaviour patterns to succeeding generations by tradition rather than by genetic mechanisms, that is, training of young, young imitating parents, conditioning or imprinting. It has been demonstrated that some parasitoid wasps lay their eggs only in the same kind of larvae as they themselves became conditioned to during their early growth and feeding. Young birds and other animals become imprinted to their own parents, to their own species, and perhaps to their proper niche, at critical stages in their development.

It is very probable that the manner in which niche segregation is passed on to succeeding generations is not the same in all species. We may believe, however, that most behaviour of animal has a genetic basis, but may become highly modified through practice, imitation and experience.

Inierspecific Competition

Segregating Effect. Charles Darwin stated in The Origin of Species the case for interspecific competition as an instrument segregating species into different niches as follows (Crombie 1947) ;

1. The reproductive capacity of organisms is greater than the carrying capacity of the environment.
2. The range of an organism's tolerance of physical conditions and choice of food is limited.
3. The failure of an organism to survive, or be born at all, may be a result of the direct action of unfavourable habitat, predators, parasites or competitors.
4. When competition occurs, it is the severest between organisms with the most similar requirements.
5. In general, the closer the taxonomic relationship between them, the more similarity there is in needs and habits of species.
6. When new forms appear in a given locality, either by evolvement there or by nvasion after evolutionary divergence elsewhere, they will either eliminate or be eliminated by their nearest relatives if they competewith them, unless.
7. Each form becomes adapted to a different niche, in which case competition between them will cease and they may occur in proximity.

Evidence that interspecific competition is the most critical factor confining a species to this or that niche is available with the expansion of the species beyond the usual limits of its niche when this competition is removed. This expansion is often evident in geographic difference in the niche characteristics of a species. In Scotland, the mountain hare occurs at high elevations, the common hare at lower ones. In Ireland, which was isolated as an island before the common hare could reach it, the mountain hare occurs at both high and low elevations and is differentiated into distinct subspecies.

The general effect of interspecific competition is restriction of a population more closely within its optimum niche. Intraspecific competition exerts pressure impelling individuals to disperse into less favourable situations. The relative pressure exerted by these two forces determines whether at any particular time the species is contracting or expanding its habitat range. Because of these interacting forces, the total population of all species in a region is constant, conditioned by the food and other resources, although the population of individual species may depend on the number of competing species composing the fauna.

Interspecific competition is reduced or eliminated altogether, when the combined requirements of all species are less than the supply of materials available. Species of land snails, insects, aquatic clams, and copepods sometimes occur together in considerable profusion but with little evidence of competition between them. Voles, during upswings in population, may become superabundant for periods of 2 or 3 years. During such periods several species of hawks and owls may feed in the same field without competing because there are more than enough voles to supply all. During downswings in the rodent population, however, competition does occur and some or all the predator species are forced to turn to the other prey.

When two species with similar niche requirements meet in competition under natural conditions, one of three things will generally happen :

1. One of the two species will be so much better adapted that it will spread rapidly through the range of the other and exterminate it.
2. One species will be better adapted to a portion of the range in which it will eliminate the other species, but the other species will be better adapted to the remainder of the range and will occupy it exclusively. Thus the two

species will occupy adjacent geographic regions with perhaps a zone of overlapping occupancy.

3. Each species will be better adapted to a different portion of the niche, to which it will become restricted, with this separation each species will then spread through the range of the other.

Predominance. It is characteristic of the taxonomic structure of communities that a few species furnish the greatest bulk of the population entirely out of proportion to the rest of the species. Thus in stream riffles, two species make up 85 per cent of the total riffles population and another two species constitute a similar percentage of the mud-bottom-pool populations. The abundant species are ordinarily of small size, herbivorous in their food habits and at the bottom of the food chains and pyramids. Predominance is better expressed in communities occurring in "harsh" environments, that is, in environments where relatively few species find conditions favourable than in those occurring in mesic ones that are favourable for many species.

Succession. Succession is the processes whereby a series of communities and habitats replace one another, beginning in an extreme habitat and ending in a mature climax ecosystem. With succession on land there is generally a progressive increase in height of the dominants, stratal differentiation of the vegetation, elaboration of internal microclimates, diversity of plant life-forms, biomass, and development of soil structure and fertility. Since the original habitat is bare of all life, at least in a primary succession, there is necessarily a progressive invasion of new species as each community develops in the sere. Species diversity tends to be greater in each successive stage, but the highest diversity index both for richness of species and equitability may sometimes occur in an intermediate or subclimax rather than the final climax stage. Successional trends may likewise vary between strata within the community.

Speciation and Cultural Evolution

Individuals within the species establish pattern of behaviour to which all adhere, so that the species functions in an efficient organized manner. Much of this behaviour is learned in each generation and thus constitutes a mores or ethics, the prototype of culture. The evolution of culture in human society is an outgrowth of these intra and interspecific interrelations in the behaviour of lower animals and developed concurrently with speciation of man.

Speciation is the process of evolutionary differentiation between populations, which may result in one older species becoming split into two or more new ones. Speciation also entails divergence of the new species into different niches or geographic areas. An understanding of the basic principles of speciation is therefore a prerequisite to an understanding of both local and geographic distribution of organisms.

Taxonomy

As the term is used by taxonomists, a population is a local aggregation of individuals that differs slightly, but characteristically, from other local aggregations of the same species. Geneticists define a population as a reproductive community sharing a common, characteristic pool. Every local population is different from every other one; but they are not easily distinguishable from one another, and therefore are given no formal nomenclature. What a population includes may vary from the entirety of a species to but a few individuals, according as the rate and extent to which individuals interdisperse between localities to make a common genepool.

Among bisexual forms, a species is a group of populations capable of successfully interbreeding and is reproductively isolated under natural conditions from other such populations, These criteria are often difficult to apply however, so reliance is commonly based on superficial or morphological characters. Species are usually morphologically distinct, but the distinguishing characteristics are sometimes barely discernible. Hybridization may occasionally occur between clearly defined species under captive or unnatural conditions, but does not occur with any significant frequency under natural conditions.

A subspecies is a geographically defined group of populations which differs in colour, size or some other taxonomic characteristic from other populations within the same species but nonetheless interbreeds with them freely, regularly and successfully, where their ranges come into contact. Subspecies are commonly distinguished by a trinomial nomenclature, although the desirability of this has been questioned. The taxonomic differences between subspecies are usually less pronounced than those between even closely related species but they are genetically fostered differences nonetheless. Race and variety are terms sometimes used as synonyms for subspecies, or for populations even less well defined than a subspecies. It is difficult to determine whether populations that differ phenotypically are subspecies or species when their ranges do not verge for in the absence of opportunity for them to commingle, it remains uncertain whether interbreeding would or could occur. The assignment of taxonomic rank under these circumstances is dependent upon considerable subjective inference. Boundaries between adjacent subspecies are frequently arbitrary in that they fit differences between the populations for some characters but not others. Differences between subspecies or populations are often correlated with differences in topography, soil, climate, or vegetation, and at least in some instances, appear to be adaptive to these differences in the environment.

Two species the distributional ranges of which do not overlap (that is, there is geographic isolation between them) are said to be allopatric. Two species are said to be sympatric when their ranges overlap, even though they may locally be ecologically segregated into different habitats; for example, the situation in which one species of snail is limited to floodplain and another species to upland habitats, or one species of rodent confined to the foothills and a closely related species to a higher zone on mountains only a short distance away. Populations should be considered sympatric if they occur within the common dispersal ranges of their young, so that there is at least possible a continuous and appreciable interbreeding and flow of genes among them.

Variation in Population Characteristics

Observable difference in structure, function and behaviour between individuals belonging to the same species are common. Actually, no two individuals, except perhaps identical twins, have exactly similar characteristics. It is the gradual accumulation of many small variations over many generations which eventually gives a population reproductive isolation and, consequently, species identity.

Non-heritable Variations

Not all variations of organisms are of direct significance in speciation. Non-heritable changes in body structures, functions and behaviour are common. If muscles are used continually and used intensively, they become thicker and stronger; if one kidney is removed, the other becomes hypertrophied; skin subjected to frequent rubbing or pressure thickens and becomes horny; and so on through a lengthy catalog. Animals progressively exposed to even more severe temperatures or lower oxygen concentrations will tolerate extreme conditions, which, had they been suddenly presented, would have been fatal. Insect larvae transferred to a new type of food often become so conditioned to it that they produce a strain that prefers food to other more usual food of the species. Chimney swifts in wilderness. Many phenotypic adaptations persist from generation to generation, either as similar responses made by each generation to constant environmental conditions; as the result of imitation of parents, conditioning of young or imprinting; or because the particular genes responsible for these characteristics have been sorted out from the general gene pool of the species. One would expect a change in behaviour or function, arisen in consequence or exposure to the new conditions, usually to presage the evolutional development of a new structure for natural selection cannot bring about the structural adaptation of perfection of an organ unless the organ is being used for the new purpose.

Polymorphism

When individuals of a population occur in more or less discrete groupings of colour phases, body sizes or other character variations that are not due to sex, age, caste, or continuous mutation, the population is said to display polymorphism. Polymorphism of a character arises in species when heterozygotes persist in an environment in the face of natural selection and homozygotes are reduced or eliminated. For instance, individuals with certain characteristics may be better adapted to environmental conditions during the spring, while individuals with other characteristics are superior in summer or autumn. This results in a mixture of types in the population more or less segregated by seasons. Likewise populations may vary in characteristic as adaptations to local habitat conditions. An increase in the frequency of a given characteristic appears to be a result of variations in the selective pressure of the environment, permitting individuals with certain characteristics of gene combinations to survive at one time or place; other individuals, at other times and in other places. In an environment that presents the same set of selective pressure year after, there commonly occurs a stability in the ratios of the different forms in which a character is manifested.

Mutations

Mutations may be the result of chemical changes in the individual gene or of chromosomal aberrations, the latter in the form of changes in the number of chromosomes (haploidy, polyploidy) or of arrangements of genes on the chromosomes (deletion, duplication, translocation, inversion). The rate at which any one gene mutates varies greatly from one kind of gene to another, but the average rate is of the order once in every 100,000 or 1,000,000 individuals. Between 0.4 and 10 per cent of the individuals in each generation may possess mutated genes. Other estimates of frequency of gene mutation are even higher.

Natural populations may, therefore, be well supplied with small. mutations of differing potential values to the organism. Mutations of different sorts apparently occur haphazardly and are not influenced by environmental conditions; only accidentally do they give special advantages to an organism. Adaptation of a species to a particular habitat or niche is effected through natural selection of the favourable mutations out of the many that occur.

The size of the population and the rate at which a particular gene mutates affect the odds that a mutation of that gene will become established in a population. However, if the gene mutates repeatedly in different individuals, the high mutation rate will greatly increase the number of individuals carrying the gene and increases the chances that the mutant character will become expressed in the population.

Natural Selection

Natural selection is a continual force exerted on each successive generation. Before natural selection can take place, however, there must be phenotypic variations between individuals from which selection can be made In order for these selected variations to have significance in speciation, they must be genetically fixed and heritable.

In most species, many more offspring are produced than can possibly survive. Because of this overproduction, there is competition between the offspring for the necessities of life, which, together with the strife between predator and prey and between organisms and their physical environment creates a struggle for existence.

There is differential survival in this struggle for existence because some individuals have structural, functional or behavioural variations that give them advantages over individuals lacking those variations The superior genotypes will make a relatively large contribution to the gene pool of the next generation. The result of differential survival and differential reproduction is popularly known as the survival of the fittest.

The accumulation of favourable variations in a population brings the species generally to a better adaptation to the physical conditions of the habitat, avoidance of predation, more efficient physiological functioning, and new behaviour patterns. Natural selection favours those variations that are adaptive, and thereby fosters the continued existence and improved reproduction of species. These adaptations may never give a perfect adjustment of the organism to its environment, since the environment seldom remains constant for any length of time. If the population undergoing the changes is geographically isolated so that the favoured changes do not spread throughout the species, then differentiation of characters leading to speciation occurs.

Natural selection is relatively less effective in small populations than in large ones. Small populations may be locally restricted, come into conflict with few competing species, and experience only a favourable habitate. Because of the low selection pressure, chance combination of genes (genetic drift) may produce characters of little or no adaptive value that yet have a good chance to persist while really adaptive characters may be lost. In populations that are increasing rapidly in size, in the upswing of cycles, say there is little selection, and non-adaptive variations may persist as well as adaptive ones. As populations come to saturate habited niches and disperse into new or less favourable habitats, competition, predation and parasitism increase, and individuals become exposed to even more severe physical and climatic conditions. Natural selection then functions and characteristics that are adaptively advantageous will tend to persist while those less favourable or even harmful are eliminated.

Mutations upon which natural selection works are often recessive. They do not become fully expressed in the phenotype except when the individual is homozygous for the character.

Nevertheless they are important, and in stabilized populations tend to persist indefinitely in constant proportion to the dominant alleles. A recessive character will become more prominent in a population if that particular gene continues to mutate toward the recessive, or if the homozygous recessive phenotype has adaptations that give it selective advantages. In this latter case, natural selection may ultimately result in complete suppression of the dominant allele so that the hitherto recessive allele becomes permanently fixed in the population as the only gene for the character.

Since emergent species usually entrain adaptations to new environments, it would appear that those characteristics by which we distinguish species and higher taxonomic categories gnenerally are such as serve some useful purpose to the organisms either structurally, functionally, or in point of behaviour. It is often very difficult to determine a useful function for all distinctive characteristics of a species, yet one can seldom be positive that a seemingly minor character does not serve, say as a releaser for some critically necessary behaviour or is not vitally important in other unsuspected ways. However, not all characters that distinguish species or higher taxonomic categories are necessarily adaptive. Some characters originally adaptive may have lost their usefulness, although they persist in the organism. With natural selection no longer acting on them, they usually retrogress and may eventually disappear, as have skin pigmentation and eyes in many cave animals, for instance. Other characters may even have a slightly unfavourable value, but be closely linked genetically with selectively preferred favourable characters and thus continue in the organism.

Cultural Evolution

Webster's dictionary defines culture as the complex of distinctive, attainments, beliefs and traditions that constitute the background of racial, religious or social groups. The culture of a species includes its social or ethical as well as its intellectual and aesthetic attainments, especially as they are symbolized or expressed in many different ways. By ethics, reference is to definite behaviour patterns concerned with interrelations between individuals that have become establishedtfor the welfare of a species. Morals refer to the ethical relations and conduct of individuals. Animals have ethics and culture in so far as they have evolved characteristics behaviour and social interrelations but culture is more complex in the higher vertebrates, particularly in man correlated with the development of a higher degree of intelligence.

Different levels of social organization may be recognized :

1. Those individuals that simply tolerate each other's presence when crowded into a restricted space.
2. Those that show some integration so that the group responds more or less as a unit, and
3. Those that have established division of labour and definite social organization.

Division of labour started evolving very early, first with differentiation of somatic and reproductive cells in colonial protozoans, then differentiation of tissues in coelenterates, differentiation of organs in vermes separation of the sexes, specialization of castes in social insects and finally segregation of species into niches in the community. Among the social insects, for instance, ants and termites have for millions of years been building elaborate dwelling places, domesticating other species, raising vegetables, storing food, practising sanitation and living under elaborate social hierarchies –analogous to what man was to evolve independently and in a different manner later on. With division of labour, each individual of species contributes to the whole, and the welfare of the social unit or community on this cooperation. These cooperating units develop high efficiency in using natural resources, with the result that the constituent individuals survive and reproduce more successfully and the species is more likely to continue to exist.

In lower animals social systems are poorly developed if one can say they occur at all. Species multiply to the limit of their food or other resources except as held in check by predators weather, disease, or decreased fecundity. With development of competition between individuals in higher species there becomes established degrees of dominance and subordination expressed in social hierarchies territorial system, and niche segregation. These form a basis for more effective community cooperation, so that food resources become divided and utilized on a continuous self-sustaining basis.

Behaviour patterns, characteristic of species, are passed from generation to generation partly through genetic mechanisms and partly by imprinting, conditioning, or training of the young to behave, in particular w ays. For instance, in birds the capacity for fear is inherited, but the young need to learn in most cases what to fear, feeding is instinctive, but the young must learn through limitation of their parents and experience what is the proper food for the species, the ability to sing is inherited, but the young must hear the song of their species to perform; and so on in the lower invertebrates, behaviour appears relatively stereotyped and automatic, but in higher vertebrates, such as fish, birds, mammals, and finally man behaviour becomes more complex with learning relatively more important. This complexity allows for more variability between the behaviour of individuals and innovations to develop faster in the mores of the species.

There can be no doubt that evolution of social systems has taken place. Probably this evolution has not been by group selection of whole systems as such, as has been suggested (Wynne-Edwards 1962), but more likely by selection of those individuals that show capacity for particular kinds of interrelations and cooperative use of resources which result in their greater survival and reproductive success. This is cultural evolution often contrasted to biological evolution concerned chiefly with structure and function rather than behaviour. Behavioural patterns, social system and community interrelations can be traced phylogenetically through animal groups in the same manner as can structural characteristics. By the same process, human ethics, culture and religious beliefs can be traced from earlier beginnings, going all the way back to lower animals. In certain respects, evolution of social systems or more among animals is comparable to evolution of vegetative systems of physiognomies plants.

We have shown how competition is of prime importance in regulation of population size segregation of species into different niches and speciation. The competitive drive for advantages is also evident in modern civilization in the development of new techniques in industry, procedures in economics, and policies in government. Nations and ideologies compete in the same manner and for the same purpose. There is selection of those techniques, procedures, policies and ideologies that are most fit and rewarding to the individual. It is difficult to imagine how competition can ever be eliminated without the advancement of civilization coming to a standstill. Man, unlike animals, may through his intelligence deliberately change the course and rate of cultural evolution and through ill-advised action reduce fitness as often as he increases it. Anthropological and historical record shows that civilizations have risen and fallen depending on the vigour and wisdom of their activities. Over the course of time the healthier, more ingenious and more intelligent civilizations have persisted longer, with successive steps leading to the complex highly developed culture of the present day.

8

Biotic Communities

Introduction

Biotic communities consist of all the plant and animal populations inhabiting a given area. They represent a higher order of biologic organization than populations, yet, since communities refer only to living organisms, they are not as inclusive as ecosystems. In other words the study of communities involves the entire biology of an area, but strictly speaking it does not involve the specific study of the interactions with the nonliving components of the environment as does ecosystem analysis. This of course, is a somewhat arbitrary distinction, and of more value for academic convenience than ecologic reality. As we shall soon see, it is impossible to talk about biotic communities without frequent reference to the nonliving environment or total ecosystem.

As with populations, it is again convenient to think of communities in terms of structure and dynamics. Structure refers to the spatial organization of communities and the way different populations are related in shape and form. Dynamics refers to the interactional processes, energetic relationships, and patterns of change within communities.

Biotic communities assume a variety of forms as diverse as life itself. If one's imagination alternately pictures a tropical forest a coral reef, a grassland, a temperate desert, an arctic slope, or the benthic depths of the ocean, one can quickly grass the tremendous variation which occurs in the tremendous variation which occurs in natural communities. Major terrestrial communities,each characterized by certain types of plants aind distinctive life forms are known as biomes. We shall consider some principles of community structure and function as they relate to a familiar biome, the temperate forest, and then to briefly, describe a few of the common community types found throughout the world.

The Temperate Forest Community

A good starting point in the study of the forest community is to examine its structure in physical and stratigraphic terms. The forest may be thought of as a layered system with each layer possessing characteristic populations and a typical organization.

The structural base of the entire community, of course, is the soil. It supplies the fundamental reservoir in terms of nutrients and water. It is both on aquifer and basic food bank. The soil itself is composed of strata or layers, known as horizons, each with definable properties. The upper layer, or A horizon, commonly called "top soil," consists of the living roots of vascular plants the mycelia of fungi, a great abundance of bacteria, protozoa, algae, and other microorganisms, and a surprising variety of burrowing animals (nematodes, molluscs, annelids, insects, various other arthropods, amphibians, reptiles, moles, shrews, etc.

Of nematodes alone, it has been said that if all plant tissues were instantly dissolved and made to vanish, the outline of trees, shrubs, grasses and herbs and their roots would still be discernible by the presence of nematodes which live within all plant tissuses.

The A horizon of soil also consists of dead and decaying organic material of plant and animal origin. The organic material is undergoing decomposition in the process of human formation of humification. The A horizon is further subdivided into various layers representing different stages of humus formation and these layers are commonly called leaf litter, duff, leaf-mold, humus, and leached zone.

The second major stratum of soil, the B horizon or subsoil, consists of a mixture of the underlying parent material and inorganic compounds which have filtered downward from the process of decomposition in the A horizon. Nitrates, phosphates, acids, bases, salts and other compounds resulting from organic decomposition have moved downward by the leaching action of water into the B horizon. The plant and animal communities of the B horizon are considerably smaller in numbers and species than those of the A horizon.

The third major layer of soil, the C horizon, consists of the basic parent material or substrate. If may be of sedimentary origin resulting from water deposition (alluvial), glacial activity (glacial till), or wind deposition (eolian or loess), or it may also be igneous, metamorphic or volcanic in origin. In general, soils forming on alluvial, glacial or eolian depositions are the richest, and some of the most productive agricultural areas of the world are found in river deltas of glaciated grasslands.

The nature of the soil profile obviously has a major influence on the animal community living within it and *vice versa*. Forest soils tend to have large populations of those organisms adapted for processes of slow organic turnover and decomposition - such organisms as fungi, isopods, millipedes, soft larval insects, and so forth. Grassland soils have higher population of those organisms which depends upon high rates or organic turnover - bactetia, nematodes, annelids and xerophilic insects.

Animal communities of soils also have major influences on the process of soil formation. Many burrowing animals play a vital role in soil aeration, physical cultivation, hydration and decomposition. Earth worms not only aerate the soil and contribute to its water holding capacities, but they bring deeper soil materials to the surface and contribute to soil mixing and decomposition. Darwin, in his classic book. The Formation of Vegetable. Mould through the Action of Worms published in 1881, estimated that earthworms brought 20 tons of soil to the surface per year. More recent studies have shown that the amount of soil brought to the surface by earthworms may very from 2.1 tons/acre/year in hot dry soils to 10.7 toh/acre/year in prairie soil, to 107 tons/acre/year in tropical delta soils (Kendeigh, 1961).

The second major stratum of soil, the B horizon or subsoil, consists of a mixture of the underlying parent material and inorganic compounds which have filtered downward from the process of decomposition in the A horizon. Nitrates, phosphates, acids, bases, salts and other compounds resulting from organic decomposition have moved downward by the leaching action of water into the B horizon. The plant and animal communities of of the B horizon are considerably smaller in numbers and species than those of the A horizon.

Other burrowing animals are also important in cultivating the soil and preventing compaction. In prairie soils, ants may bring over a ton of the surface per acre per year, and prairie dogs may excavate 30 to 40 tons per acre in and around their burrows. Not only do all burrowing animals till the soil physically, but they contribute to its chemical maturation by processing organic material through their digestive systems.

Soil formation is a slow process of ecologic maturation. Forest and grassland soils involve many centuries of organic accumulation, and they represent the stored capital of very long-term ecologic processes. The rate at which these processes occur varies greatly according to climate, topography, substrate, biologic community and all those ecologic factors influencing the processes themselves. It is thus impossible to give a precise figure on the rate of soil formation. Some temperate soils are literally millions of years old. In fact, several studies have shown that temperate soils require 10,000 years to develop recognizable soils profiles and some profiles and some profiles are not really well-developed in less than 200,000 years. Tropical soils develop much more rapidly, of course, but they are thin and susceptible to desturction. It is thus important for modern man to realize that soild is not a synthetic commodity, or one that can be manufactured in a short time.

In proceeding upward from the fioor of the forest, the biotic community thing out to a certain extent, animals become more widely spaced in three dimensions, and they become more mobile. The plant community is dominated by herbs and shrubs, and the animal community by insects, birds and mammals. Mammals, including deer, rabbits, mice, squirrels, shrews, racoon, possum, etc. actively forage through the lower layer of this comunity.

Many animal taxa including annelids, some molluscs, myriapods and solid-welling arthropods do not enter this realm and are seldom if ever found above the surface of the ground. There are exceptions, of course such as certain snails (molluscs) which climb trees.

Species Diversity

In 1958 the eminent ecologist of Yale university, G.E. Hutchinson, delivered an address of the annual meeting of the American Society of Naturalists and used the title, "Homage to Santa Rosalia or Why Are There So Many Kinds of Animals ?" He took the title from the sicilian saint, Santa Rosalia, near whose sanctuary on Monte Pellegrino, Sicily, Hutchinson had found a small pond rich in animal life and limestone cavers rich with the bones of pleistocene animals. The setting gave Hutchinson addittional inspiration to reflect on the ecology of evolution. Why has a natural selection produced such a great variety of living organisms - over a million known species of animals alone, three-quarters of which are insects ?

This topic has been a major question in ecologic and evolutionary theory for many years. In discussing the evolution of species diversity, Hutchinson emphasized the mosaic nature of the environment.

When the environment is viewed in very small units of space, it may be considered an infinitely variable mosaic of different conditions. Hutchinson pointed out that the small physical size of many animals, such as insects, has permitted them "to become specialized to the conditions offered by small diversified elements of the environmental mosaic, and this clearly makes possible a degree of diversity quite unknown among groups of larger animals." Furthermore, the addition of each new plant or animal to the community increases its diversity, thus creating new conditions on which natural selection can operate. Diversity, then, becomes a self-stimulating phenomenon.

The limiting factors on diversity are the severi y of the physical circumstances to which life forms can adjust. All life having a common biochemical basis, has certain central tendencies in the conditions around which life processes operate with the greatest efficiency. This does not infer thatt all species of plants and animals have the same environmental requirements but it does mean that more plant and animal species are going to survive and thrive of environmental conditions approaching the normal requirements for protoplasmic processes. In other words, the biochemical unity of life insures that more plant and animal species are going to do well at 80. to 90. F (27.. to 32 . C) than at 32...F (0 C). Plant and animals can, of course, adapt to the latter but the selection for this will be substantially more rigorous, and the available niches much fewer.

Seen in this light, it becomes reasonable to theorize that those environments whose physical circumstances are harshest and most inimical to life will have the least diversity. Thus we would expect desert, ocean depths (below the light zone). polar regions and snow-capped mountains to have the least diversity in their biotic communities. This is indeed, exactly true. These are biotic communities characterized by the lowest numbers of different kinds of plants and animals, though there may be very large populations of these few species.

On the other hand, we would expect those environments with conditions most nearly approaching the ideal for the processes, that is, warm temperature, ample light and plenty of moisture, to have the greatest diversity of plant and animal life. This also is true. The tropical forests are far richer in number of species of plants and animals than any other biotic community on earth.

The relationships between species diversity and community stability have been mentioned previously. In general, the greater the diversity, the greater the stability. Communities with many species have a more complex web of food relationships, and each species has greater diversity in its food and cover resources. Thus each species has alternatives in meeting its life requirements. If has various ways of meeting new environmental shortages and contingencies. There is in other words, a cushioning network of checks and balances. It must be remembered that we are speaking in broad terms and there are undoubtedly many exceptions to hese generalities. Nonetheless, these generalizations are valid as overall principles of community ecology. As a result, any major change in a single plant or animal population in a polar community is likely to have a far greater impact on the total community than similar change in a tropical community. It thus becomes important to think of communities in terms of their vulnerability to ecologic upset if change occurs in any one of their member populations.

Succession

Biotic communities change with time as their plant and animal communities change. This process is known as a succession, and it involves a sequence of community types from pioneer stages to mature or climax stages Each community in the series is known as a seral stage. One of the best ways to appreciate succession is to consider the sequential growth and development of biotic communities on a cleared forest.

If a forest is cut, and the land cleared to the soil, a succession of plants will invade and grow on the exposed surface. The first plants which invade are typically those capable of seeding in quickly on disturbed land or capable of generating from viable roots left within the soil. This depends not only upon the former vegetation, but also upon seeds from surrounding plant communities, as well as the characteristics of the soil and climate. Typical invaders in the United States are herbaceous weeds such as ragweed (Ambrosia sp.), lambs quarters (Chenopodium album) T dock (Rumex sp.), horse weed (Erigeron canadensis), plantain (Plantago sp), crab grass (Digitaria sp. et al) and woody vines such as honeysuckle (Lonicera sp.) and poison ivy (Rhus taxico-dendron). As soon as the first plant germinates and becomes established, the community increases vastly in physical complexity. A surface of bare earth now has shaded areas which differ in light, moisture and humidity. This enables other seeds to grow, and new species become established. Woody plants appear, again those capable of withstanding disturbed environments and harsh conditions, such as black locust (Robinia sp.), sumac (Rhus sp.) and howthorn (Crataegus sp.). Pioneer animals also arrive, including ants, beetles and flying insects. Birds begin to forage over the new community searching for seeds and insects. Small mammals may venture forth from adjacent forests or grasslands Lack of animal traversing the area adds organic nutrients. Within a year or two a surprisingly complex community has arisen, and is growing rapidly. As plants grow, they create more shade, and thus continuously modify the light and moisture conditions on the surface of the ground. These initial pioneers which thrived on total sun exposure are now less favoured and more shade-tolerant seeds can germinate more successfully. Woody shrubs emerge above the herbaceous layer and compete more successfully for the ambient light.

The shade developing beneath the herbaceous and shrubby layers permits the growth of other tree species such as maples, elms or oaks which could not tolerate the initial exposure of sun and wind. This represents the first beginning of the forest. Further succession continues to involve all the dynamic processes of plant and animal competition. Over a period of years, the initial invaders and pioneers begin to drop out. Trees nature first the fast growing locust and sumac, but as they reach the limits of their growth and their relatively short life-spans the slower growing oaks (Quercus sp.), maples (Acer sp.) and elms (Ulmus sp.), maples (Acer sp.) and elms (Ulmus sp.) begin to take over. The forest may now be 40 to 50 years old, reaching a height of 50 to 60 feet.

The ground cover and intermediate layers have now matured to include those shade-tolerant species capable of living in the forest under study. The insect and bird populations have changed from open field forms to forest species. By 70 to 80 years, the forest may be approaching relative maturity, with tall trees in the 60 to 80 foot range, deep shade and cool, moist conditions on the forest floor capable of supporting the rich array of life described earlier in this chapter. The climax first may be approached in 80 to 100 years, when it becomes a stable and self-perpetuating community. Processes of change still occur. Old forest trees die and crash to the ground, creating a microcosm of renewed successional development, but the main forest continues in equilibrium as a permanent community as long as environmental conditions remain favourable. If it is cut, the process of succession begins again. If it is defoliated, flooded, burned or dehydrated, it will change and attempt to seek a new form of equilibrium with its environment. The tree species that finally emerge as the dominant-members of the climax community vary according to the climate, topography and soil type.

It would be incorrect to assert that the process of succession is always consistent and predictable. It may take a variety of forms, and may become arrested at any of several stages. Early in the shurb stage, the biotic community may become overwhelmed by excessive growthof woody vines such as honey-suckle or poison ivy, so that trees cannot develop adequately, or it may stay in the herb or shrub stage if the land is subjected to continuing disturbance by fire, pollution, or and scarring. Throughout the condtryside, especially around cities, one can find many examples of disturbed and unattractive vegetation that never gets out of the weed, shrub ar vine stage.

The above discussion has mainly concerned a deciduous forest community, but a similar account could be given of coniferous forests, and, in fact, of all biotic communities. Thus, a prairie grassland, a tropical savannah, a desert arroyo, an estuarine marsh, a cedar bog, a pond, a lake, a delta, a mangrove swamp, or a coral reef all have characteristic patterns of development and succession. In some communities, successional development may involve hundreds or even thousands of years before mature climax conditions are achieved. Primary tropical rain forests, for example, are the culmination of thousands of years of successional development. Once cut, they are not recreated in a few decades or a hundred years.

It should be emphasized that succession is also a naturally occurring process and is not confined to disturbed habitats. Theoretically, each biotic cammunity tends toward maturation and stable climax conditions, but there is sufficient natural change in ecosystems so that succession is a continually recurring process in many situations. The grassland and forest are subjected to natural fire as an ecologic process, the desert arroyo to flash flooding, the marsh, bog, pond, lake, delta and swamp to filling, and the coral reef to the continual processes of construction and destruction. Human activities often increase and accelerate these processes far beyond the natural rates of occurrences. In so doing, man impedes the development of maturity and ecologic stability in biotic communities.

An example of natural succession is the filling of a small pond and its eventual conversion into a terrestrial community.

As the pond ages, it becomes more filled with aquatic vegetation silt. Emergent vegetation surrounds the edge creating a marsh community. Accumulations of organic material and sedimentary deposits fill in the water volume, soil is formed at the edge, and terrestrial grasses and herbs become established. Eventually, shrubs and trees complete the process. Man accelerates this process by increasing the siltation process and enriching the waters with nutrients so that more extensive plant growth occurs. The process can be retarded by dredging the pond or manipulating the water level to reduce aquatic vegetation. Lakes proceed through some of the same stages, but if sufficiently deep, a point is reached in lakes where the scouring action of currents slows or even balances the siltation and sedimentation rate.

It is important for man to understand successional processes so that he can manage biotic communities for maximum productivity and aesthetic concerns. In agricultural lands and harvested forests, man has learned to maximize productivity for his own economic gain. This is done by artificially maintaining communities in early successional stages, where productivity is greatest, and by channelling all possible production into the species to be harvested. That is, man maintains a cornfield or a wheatfield in an early productive stage by excluding all other pioneer species. He maintains a forest in an early successional stage and directs all productivity into the species of tree he wishes to harvest. This is all well and good provided the cost of such management is not too high in terms of toxic chemicals that must be applied to eliminiate other species, or in terms of chemical fertilizers that must be applied to support an artificial single-species community.

Quite another principle prevails in the management of parks and sanctuaries where man wishes to maintain balanced plant and animal communities for esthetic and scientific values. Here it becomes essential to understand and promote natural succession so that mature and stable communities are achieved.

In this day of economic pressures it is increasingly difficult for man to appreciate the value of natural biotic communities. What good is a mature forest if it is not harvested, what good is an estuarine marsh, what good is a relic prairie ? These are all questions being asked in greater frequency by citizens focusing on economic progress. The challenge the ecologist to explain the message of ecology in factual terms, and to emphasize the importance of balanced biotic communities to the total life support capabilities of the world. The answers are not easy, and are not always available. Much remains to be learned about the full range of functions of an estuarine marsh, for example, in maintaining the productivity of the oceans. This represents one of the greatest current challenges for ecology; namely, to understand and explain in a meaningful way, without great fanfare and unreasonable emotion, the importance and value of diverse biotic communities. Otherwise economic man will see little reason why he cannot continue to convert the entire world to paved streets, grass lawns, grain fields, and greenhouses.

Major Biotic Communities

It is worthwhile at this stage to take a brief look at several different communities and highlight some of their outstanding features. This has been done in various degrees of detail by a variety of authors, and no attempt will be made here to be comprehensive or thorough. Descriptions are offered as capsule summaries of six interesting and representative communities; the arctic tundra, northern coniferous forest, desert, grasslands, estuaries and tropical rain forests. This limited list unfortunately omits many other important and interesting communities : savannah, chapparal, temporate rain forests tropical deciduous forests, fresh-water swamps and marshes, lakes, rivers, coral reefs and open oceans. Volumes could be written about each of these, and many others.

Arctic Tundra

Arctic tundra is the biotic community occupying the northern latitudes, generally above 60 north latitude. The dominant plants are lichens (the combination of algae and fungi known as "Reindeer moss"), arctic grasses and sedges, and some dwarf shrubs, especially dwarf willow (Salix sp) and cranberry (Vaccinium sp.). During the brief summer the flowering season is abrupt and colourful. The vegetation is underlain by chick spongy mass of undecayed deed vegetation. Bacterial action is slow due to the low temperatures, and the substrate is often permanently frozen, a condition known as permafrost.

The animal community is simplified into relatively few species of birds, mammals and flying insects. The dominant mammals are the caribou (Rangifer arcticus) or reindeer (Rangifer tarandus, a domesticated species of caribou), arctic hare (Lepus arcticus), lemming (Lemmus sp. or Dicrostanyx sp.), and a arctic fox (Alopex lagopus). The dominant birds are ptarmigan (Lagopus lagopus, a gollinaceous grouse-like bird), the longspur (Calcarius sp., a sparrow-like bird), and the snowy owl (Nyctea nyctea). Great numbers of migratory waterfowl (ducks, geese, swans and shorebirds) invade the tundra in the summer for breeding. Insects are dominated by the biting diptera, blackflies and mosquitos.

Northern Coniferous Forest

This is a forest belt at latitudes below tundra, occupying major portions of Alaska, Canada, Scandinavia and Siberia, generally between 50 and 60... north latitude, but with extensions in mountainous regions further south. The plant community is dominated by spruces, firs, pines and hemlock, with a ground cover of mosses, grasses, sedges and cold adapted herbs. Decomposition is slow, as in the tundra, and the forest floor tends to accumulate a layer of dead organic material with excellent waterholding capacity.

Animal populations are more varied than in the tundra, but are still characterized by great seasonal change and wide population fluctuations. Typical mammals are the snowshoe hare (Lepus americanus), lynx (Lynx canadensis), squirrels (Sciurus sp, Tamiasciuruss sp.), marten (Martes americana), mink (Mustela vison), fisher (Martes pennanti), wolverine (Gulo luscus), woodland caribou (Rangifer caribou), deer (Odocoiteus sp.), moose (Alces americana) and black bear (Ursus americana). Typical birds are ruffed grouse (Bonasa umbellus), crossbills (Loxia sp.), siskins (Spinus pinus) and a multitude of other passerines which migrate in for summer breeding. Winged diptera, such as blackflies and mosquitos, beetles and moths are among the dominant insects.

Grasslands

The grassland community covers vast stretches of land within continental interiors. These are areas characterized by plains and 10 to 30 inches of rainfall, per year, generally less than in most forested regions.

The plant community is dominated by grasses, but it contains a great variety of other herbs and forbs as well, particularly legumes such as lupine (Lupinus sp.), trefoil (Trifolium sp.), clover (Petalostemum sp.), and vetch (Vicia sp.); composites, such as aster (Aster sp.), firewood (Epilobium sp.), and daisy (Chrysanthemum sp. or Bellis sp.), and members of the Ronucutacead, buttercups (Ranunculus sp.), anemone (Anemone sp.), columbine (Aquilegia sp.) and larkspur (Delphini sp.). In all a grassland or praairie floro may contain over 120 species in more than 10 families. The dominant green grasses are predominantly or three general sizes : tall grass, often 6 feet in height, which includes tall blue stem (Andropogon gerardi), switch grass (Panicum sp.) and Indian grass (Sorghastrum sp.); medium grass, 2 to 4 feet in height, which includes little blue stem (Andropogon scoparius), feather grass (Stipa short grass) eլss than 2 feet in height; which includes buffalo grass (Buchloe sp.), blue grass (Poa sp.) and blue grama or mesquite grass (Bouteloua sp.).

As previously mentioned, the annual production of organic material in grasslands is great, organic accumulation is rapid, and a thick layer of humus is produced. Grasslands soils are among the thickest and richest in the world. Roots of the grasses and forbs generate up to 6 feet in these rich soils.

The animal populations of grasslands are also rich and varied. Among the mammals, the ungulates and rodents have flourished with great populations of bison (Bison bison), antelope (Antilocapra americaana), prairie dogs (Cynomys sp.) ground squirrels (Citellus sp.) and gephers (Geomys bursarius), and the mammalian predators are represented by coyotes (Canis latrans), foxes (Vulpes sp.) and badpgers (Taxideaa taxuss). Typical birds are prairie chickens (Tympanuchus cupido), sharp-tailed grouse (Pedioecetes phasianellus), meadow larks (Sturnella sp.) and a variety of sparrows. Insects are abundant, with grass-hoppers and locusts being among the most conspicuous.

Deserts

Deserts are arid biomes with usually less than 10 inches of rainfall per year. Deserts usually occur in areas of high pressure (Sahara, Australian), rain shadows of mountains (Mohave, Iranian) or high altitudes (Tibetan, Gobi and Bolivian).

Some of the direct deserts may not have rain for a period of many years. One such area in Chile has not received precipitation for over 20 years. Such deserts may be virtually devoid of living organisms over vast areas. In the Sahara desert of southern Libya, one can travel for hundreds of miles without seeing of living plant or any form of life. Most deserts, however, do have some water resources from occasional rains or from ground water, and they possess a considerable variety of living organisms dependent upon these water resources.

The predominant plants of deserts are succulent species with waxy surfaces, such as cacti, which can conserve water for long periods of time, or deciduous shrubs, also with thick waxy leaves. The growing season is very short, and the flowering period amazingly abrupt, occurring suddenly after flash rains. Many deserts have little or no plant life over extensive areas of sand and rock.

Estuaries

Estuaries are bodies of water where fresh water from the land mixes with and measurably dilutes sea water. They represent dynamic combinations of fresh water and marine communities and they have a number of unique characteristics. They are particularly important to man by virtue of these unique biological attributes.

Several different types of estuaries exist, representing various topographic formations and geologic histories. Some develop from coastal submergence and drowned river mouth such as Chesapeake Bay and Delaware Bay. Other develop as coastal lagons behind barrier sounds, such as Albemarie, Pamlico, and Currituck sounds inside Cope Hatteras. Others arise as steep glaciated fjords such as those along the coasts of Norway, Chile and British Columbia. Finally, many develop within great delta systems such as those of the Nile, Mississippi, Mekong and Ganges rivers.

Tropical Rain Forests

The tropical rain forest is a biotic community which has always stirred the human imagination, and rightly so. It is one of the most dramatic and awesome biologic systems on earth. Such forests occur within the Tropic of Cancer and the Tropic of Capricorn (23° 27' N and S latitude), in areas with more than 80 inches of rainfall per year. Usually such forests have one or more dry seasons per year in which less than 5 inches of rain fall per month. The monsoon forests of southeast Asia are one example, and although it rains throughout the year, concentrations of rainfall occur twice a year in the monsoon periods.

An important characteristic of the tropical rain forest is its relative climatic stability. The temperature variation between seasons is light and usually less than the temperature variation between day and night, or between different microhabitats within the forest. For example, in Singapore at the tip of the Malay Penninsula the diurnal variation in temperature is often 20°F (from 75 to 95°F), whereas the variation in temperature from January mean to the July mean is only 2.5° (from 87° to 89.5°F). By contrast, the variation from January to July mean in Dawson City, Alaska is 81.7°F (from -15°F to nearly 70°F).

9

Natural Resources

Introduction

Whether we run out of a particular resource depends in part on how we define resource. In broad terms, a resource, or natural resource, is anything needed by an organism or group of organisms. In other words, a resource is something useful but for humanity what is useful or useless can change because of technology, economics, and the environmental effects of getting and using a resources.

Technology cannot bring back an extinct animal resource or a paved over wilderness area, but it can extend the supply of some resources by improving them, using them more efficiently, or recycling them. For example, a unit of today's steel provides 43% more structural support that it did a decade ago. Today we get 7 times more electrical power from 1 ton of coal than we did in 1900. Similarly the energy needed to produce 1 ton of pig iron has fallen eight fold since 1800. In the United States in 1900, only 10% of the copper was recycled. Today about 40% is recycled. However, while many matter resources, such as copper, lead and silver, can be recycled, we can never recycle energy resources. Once a fossil fuel resource, such as coal, oil or natural gas, is burned, it is gone forever as a useful energy source. The concentrated energy in the fossil fuel is released as heat, which is eventually dispersed into the earth's atmosphere. From there it flows back into space.

Sometimes technology can solve the problem of a scarce resource by finding a substitute or replacement. For instance, as a structural material bronze replaced stone, iron replaced bronze, steel replaced iron, and now aluminium and reinforced plastics are replacing steel for some structural uses. As an energy source, animals replaced human muscle power, coal (to produce steam) replaced animals, and oil replaced coal for many uses. There is a vigorous debate over whether a combination of coal derived energy and nuclear power (obtained from Uranium) or a combination of solar, wind and, and plant (bio mass) energy may soon replace petroleum and natural gas.

In addition to technology, resource use is tied to economics. Something is useful as a resource only if it can be made available at a reasonable cost. For example, once we deplete the easily available supplies of a resource, we have to look harder and dig deeper to find remaining supplies. If the costs of findings and making a scarce resource available rise, the resource will eventually become too expensive for most people. Higher costs may stimulate a new search for new supplies or make mining and processing lower grade deposits economically feasible. But regardless of what we are willing and able to pay, we can't get a resource out of earth if it is not there.

There can also be an economic limit to recycling. Typically, recycling is cheaper than mining virgin materials, but only if the material to be recycled is not too widely dispersed. For example, if products made from iron and steel, such as cars and toys, are thrown away, buried, and widely scattered through use, labour and energy costs for finding and collecting the objects may be too expensive to make recycling feasible.

The continued use of a resource may also depend on the impact its mining, processing, and use has on the environment. Even if affordable supplies of a resource are available, its use (at least for certain purposes) may have to be abandoned if this use seriously threatens human and other forms of life. Sometimes the environmental effects of a resource use can be minimized and cleaned up. But this costs money, and in some cases the costs may be so high that we can no longer afford to use the resource.

Natural Resources : Definition

The sum of all physical, chemical, biological and social factors which compose the surroundings of man is referred to as environment and each element of these surroundings constitutes a resource on which man draws in order to develop a better life. Thus, any part of our natural environment - such as land, water, air, minerals, forest, rangeland, wildlife. fish, or even human population – that man can utilize to promote his welfare may be regarded as a natural resource.

Ramade (1984) defined a resource as a form of energy and/or matter which is essential for the functioning of organisms, populations, and ecosystems. In the particular case of humans, a resource, in his words, is any form of energy or matter essential for the fulfilment of physiological, socio-economic, and cultural needs, both at the individual level and that of the community.

The functioning of ecosystem or of technological civilization involves a continuous flow and consumption of energy sun being the source of it as regards ecosystem and fuels as regards technological civilization. In the light of this fact, a resource may be defined as one of the various forms of primary energy present in the nature.

The five basic ecological variables – energy, matter, space, time, and diversity are sometimes combinedly called natural resources. As such, laws which govern changes in these quantities have great applicability in the resource use.

Classification of Natural Resources

The classic subdivisions of resources include renewable and non-renewable resources. Renewable resources are those which are reproducible and are obtained from the biomass of living organisms, *e.g.*, forests and fish. Resources which are not reproducible and are obt ined from the finite non-living reserves are called non-renewable resources, *e.g.*, coal and metals. Solar energy although having a finite life, as a special case, is considered as a renewable resource in an much as solar 'stocks' are inexhaustible on the human scale.

Some authors prefer to classify resources into biotic (or living) and abiotic (or non-living) resources, forest, agriculture, fish, and wildlife falling in the first category and land, water minerals etc., in the second.

A broader classification of resources, as proposed by Owen (1971), is represented in Fig. 1. Quantity, mutability, and reusability are the attributes on which this classification is based. Inexhaustible and exhaustible are the main categories of resources based on their stocks in nature. Inexhaustible resources are further divided into two groups - Immutable and mutable resources, depending upon the possibility of their qualitative degradation as a result of man's activity. On the other hand, exhaustible resources, being limited in occurrence, are vulnerable to both quantitative and qualitative degradation. Thus, their availability depends upon the method of use, so these are classed into maintainable and non-maintainable resources. Supply of maintainable resources could be made to last for long through wise use, however, some of them have capacity to reproduce - called renewable, while other cannot reproduce - called non-renewable resources. Non-maintainable resources have static supply and when destroyed or consumptively used are irreplaceable, yet some of them have potentiality for reuse - called reusable resources, and rest whose permanent exhaustion is certain are called non-reusable resources.

Conservation and Management of Natural Resources

The term 'conservation' to many means 'hoarding' or to control supply of goods in such a way that some part is left for the nature. Most people interpret conservation as total protection or restriction on consumptive use of resources, therefore, wildlife conservation to them means a total ban on killing of any animal and the forest conservation means a complete moratorium on all fellings the term, in fact, is derived from two Latin words—con (=together) and servare (=to keep or guard), thus meaning 'to keep together'. The dictionary meaning of the word is 'an act of preservation'. The term was proposed by Gifford Pinchot in 1908 (see Smith, 1958) who conceived it from the title of 'conservarcs' given to British officials in India appointed to look after the protection of natural resources.

Prof. H.M. Rose defined conservation as "the optimum allocation of natural, human, and cultural resources in the scheme of national development, whereby maximum economic and social security will be assured" (see Smith, 1958). In 1967, the American President John F. Kennedy looked upon conservation as "the wise use of our natural environment the prevention of waste and despoilment while preserving, improving and renewing the quality and usefulness of all our resources". World Conservation Strategy defines conservations as "the management of human use of the biosphere so that it may yield the greatest sustainable benefits to present generation, while maintaining its potential to meet the needs and aspirations of future generations". Thus, the conservation is a practice embracing preservation, maintenance, sustainable utilization, restoration, and enhancement of the natural environment. Summarily, the true aim of conservation is two fold : (i) preservation of quality of environment, and (ii) to ensure continuous yield of useful material-living or non-living by establishing a balanced cycle of harvest and renewal.

The concept of resource conservation has now come under a still wider field of environmental management which is, in essence, the process of allocating natural and man made

NATURAL RESOURCES

- **Inexhaustible**

  Unlimited or unending. Quality may be degraded, not quantity.

  - **Immutable**

    Quality not degraded by men's activity, *e.g.*, atomic energy, wind power, precipitation, tidal power.

  - **Mutable**

    (Misusable)

    Quality may be degraded by man's activity, *e.g.*, solar radiation, water, hydropower of running waters.

- **Exhaustible**

  Limited. Both qualitative and quantitative degradation possible.

  - **Maintainable**

    Availability depends upon method of use by man.

  - **Nonmaintainable**

    Total quantity static, when consumptively used, not replaced.

Renewable

Perpetual harvest possible, *e.g.*, fertility of soil; bio-products of land, fresh water and sea; human power.

Nonrenewable

Once gone, no replacement, *e.g.*, species of wildlife

Reusable

Consumptive use small, *e.g.*, germs, non-consumptively used metals like gold, platinum, silver, iron, copper, zine.

Nonreusable

Consumptive use high, *e.g.*, fossil fuel; non-metallic minerals like gypsum, sand, salt; consumptively used metals like lead in gasoline and paints, zine in galvanized iron, tin in tooth-paste containers.

Fig. 1. Schematic representation of classification of natural resources.

resources so as to make optimum use of the environment in satisfying basic human needs at the minimum and more, as far as possible, for an indefinite future, and at the same time preventing depletion and degradation of resources.

Environmental management being a too vague, general, or trivial subject has not developed as a full-fledged discipline so far, further, in many developing countries it is viewed as being irrelevant and counter-productive, while in advanced industrialized countries it is often interpreted as a means of coping with the 'burdenl of environmental legislation and problems put forth by 'over-zealous' environmentalists; what to talk of under-developed world ? The global population, however, cannot and should not overlook and deny the fast deterioration of environment world over resulting in dwindling natural resources, pollution of air, water and land, extermination of a number of life forms, down going health status, danger to cultural assets, and many socio-economic backlashes. The root of the problem lies in too many people using too many resources wastefully.

Since the total natural resources available are limited, the wayout to the problem seems only through control of population and also the per capital consumption of resources. Because it is difficult to bring down per capital consumption, the recycling and stringent conservation can bring hope for sustained development without reducing per capital use of resources.

Land

Soil consists of the weathered layer of the earth's crust with living organisms and products of their decay intermingled. Lying on the rocky core of the earth, this complex mixture is one of the four prime requisites for life. Along with sunlight, air, and water, soil supports all living beings on land, plants as well as animals. All higher terrestrial plants have their roots in soil from which they absorb life sustaining water and nutrients. Man and all land animals, in turn, depend on these plants for food, directly or indirectly. Foods, vegetable oils and fats, forage for livestock, fibre, wood and various other indispensable agricultural products come from soil. Thus, a nation's economic well being is inextricably linked with the fertility and abundance of her soil resources. Productive land is since the source of human sustenance and security everywhere at all times, this resource because of mounting demands of swelling population and long mismanagement would put in jeopardy the very survival of man.

Land Use and Abuse

The land although appears to be resource available unlimited but, in fact, its injudicious use would limit the availability of this indispensable life-support system. Thus, a land use may be poor or good. If in a region of law rainfall grasslands are replaced by wheat crops, a temporary desert will eventually result; or if the human habitation and factories are developed on flood plains, the damage is almost inevitable. These are the instances of poor land use. Conversely, if the grasslands are used for moderate grazing by livestock, and if the flood plains are used for forestry or agriculture, these will imply good land use. Ecological principles have great applicability in land use planning. Classification of use should, therefore be based on natural ecological features of the land in question, *viz.*, soil type, slope, biotic communities etc.

Rural Land-Use Planning

To a farmer, soil is the essence of survival and his economic well being is firmly linked with the fertility of his land. In rural land-use planning, concentration is chiefly devoted to creating and developing more farmland by removing forests and natural grasslands, draining marshes and swamps, channeling streams for irrigation, and so on. Unfortunately, no effort is made to save existing prime farmland from degradation by ill-planned development. Rather than exclusive farm, range, or forestry oriented planning, the land use should be made under broad consideration of social, economic, and ecological intricacies.

Urban Land-Use Planning

Deteriorating quality of urban and suburban environment is to a great extent the result of injudicious land use and is a threat to the whole socio-economic system. Thus, planned cities are as necessary as planned farms. Although, the implementation of land-use plans is infinitely more difficult in urban areas than In rural ones because of the human problems involved, yet if not heeded it would lead to chronic pollution, epidemics, and social disorder. In any urban land-use planning it should be seen that : (i) at least a third of all land remains as protective open space in the form of parks, green belts, or refuges etc., including adequate space for efficient treatment of domestic and industrial wastes; (ii) residential housing around villages or cities should be developed in clusters with each unit separated by broad green belts; (iii) typical topographic situations like stream valleys, steep slopes, lakes, and marshes etc., should be retained free from houses and other high-density uses; and (iv) waste disposal areas and industries should be kept apart from human habitations.

Management of Land

From the above account it is clear that major problems emerging out of abuse of land in cities include unsafe and unhygienic dwellings and those in rural areas are of erosion of soil and loss of fertility. While urban problems can be prevented and cured by proper town planning and municipal facilities, the problems relating to rural areas need long term soil conservation programme.

Prevention and control of soil erosion. The primary purpose of soil conservation is to prevent soil erosion and heal the damage where it has not advanced too far to respond to curative methods. This is possible largely through two means :

1. The land should wear a vegetational cover throughout the year. Plan roots bind the topsoil and prevent it from being drifted so easily with water or air. The vegetative cover also slows up runoff water and causes much of it to percolate into ground that promotes more vegetative growth. The vegetation – whether grasses, trees, or cultivated crops-should fit to the capabilities of the soil and water available.

2. Engineering and agronomic practices should be applied conjointly to counteract the erosive force of both water and wind on the soil. These include contour cultivation, cover crops and crop rotation, and contour stripcropping, used alone or in combination with terracing where it is required. When terraces are properly constructed and adequately supported by approved cropping and tillage practices, they prove to be one of the most effective erosion-control measures applicable to cultivated sloping lands.

Reclamation of eroded lands. Conservationists are divided into two camps over the issue of reclamation of land already arrested by ravines. One camp of conservationists and naturalists believes in the policy of 'forgive and forget' and advocates that ravines should be allowed to exist, while the other feels that efforts should be made to regain the lost land through bulldozing or some other means. If the ravines are left in there present state, these will spread further and engulf more of productive land, while the violent means to regain the ravines in their near original condition may provoke numerous agencies of nature to take revange on man. Thus, following considerations are important in negotiating the eroded land :

(i) Ravines should be provided with sufficient and suitable vegetative cover.

(ii) Instead of agriculture, these lands should be reclaimed for forestry, pasture, or horticulture.

(iii) Their deficiency in nutrients and moisture for plant growth should be improved.

(iv) Further misuse of such land shouid be prevented. Overtrempling by man and cattle shohld be avoided.

(v) Post damages should be repaired and recompensed under strict ecological consideration. For instance depressions may be filled up, slopes easened out, cattle trails fenced etc.

(vi) Vegetative cover provided should be protected against reckless destruction by local population.

Water

The origin of water on the earth is not clear so far. However, the current presumption is that the primordial earth had no oceans, and perhaps very little atmosphere. It is believed that the volatile constituents bound in the earth's crust, oozing the surface through volcanos, rock movements, and hot spring, condensed to form the ocean and the atmosphere. This way, perhaps this remarkable combination of hydrogen and oxygen – called water – came into being and eventually became an indispensable component of the earth's environment.

Water moves from the ocean to air to land to ocean in a set cyclic pattern called hydrologic cycle. In other words, this cycle represents a continuous process of exchange of water from the earth's surface to the atmosphere and *vice versa* under the influence of solar head.

It is estimated that of the total available water 97.3 per cent is contained in the oceans and out of remaining 2.7 per cent most is in solid from. The amount of water actually available in the atmospher is a very minute fraction and estimated to be only 1 × 10 – 5 of the world's total water.

The factors which control the process of evaporation and evapotranspiration include temperature, humidity, and wind. It is a continuous process going on day and night from the ocean surface, ground surface in land waters (like lakes and streams), plant and animal surfaces, Rising in the atmosphere this water forms clouds that float around in gaseous or droplet form. The atmosphere above the earth up to varying heights of 10 to 17 km (depending upon latitude and season) is called the troposhere up to which the clouds float and in which all weather changes take place. Water returns to the earth in the form of precipitation, most of which takes place over the ocean and a little on the land surface. Rainfall on the land surface may have following fates : it may percolate through the soil and become a part of ground water regime; accumulate in a pond, lake or reservoir as inland surface water; flow down to the ocean in the form of stream; taken up by plants and animals; or evaporate.

Water Requirements and Uses

Goethe rightly said that, "Everything originated in the water, and everything is sustained by water". Water is needed to fulfill diverse requlrements in so many diverse ways. It is vital to life since for all physiological activities of plants and animals it is essential. On an average it constitutes 80 per cent of protoplasm ranging from 8 per cent in dry dormant seeds to 95 per cent in jelly fish. It serves not only to quench the thirst but also to meet the food requirements because it is an essential raw material in the process of phtosynthesis through which green plants make food that is used by all trophic levels directly or indirectly. Ambient humidity which is indeed the invisible form of water that surrounds us is necessary to prevent dessication of terrestrial life forms. Thus, it can be said that water is a biological essential and it serves as milleu internal as well as milieu internal.

Flood plants of rivers have been the cradle of civilization and centres of population since ancient times. The reason being their high fertility and easy accessibility to water for various uses like drinking, washing, and transport. Egypt was developed on the flood plains of the Nile; Babylon on the flood plains of the Tigris; and North India on the flood plains of the Ganges. Even today the Mississippi plain is about the most densely populated area in the USA as flood plains of the Rhine in Europe, and the Yellow river in China.

Besides temperature, water is the other such key factor that influences the global ecology. Covering about 70 per cent of land surface and circulating in hydrologic cycle water influences weather and climate of any region and thus its flora and fauna.

Agriculture is dependent on water, may it be rainfall, or surface or ground water irrigation. It eventually is a more important commodity in a country like India because her largely agrarian economy depends critically on water. Huge amount of protein rich food in the form of fish, shell fish, and prown is obtained from the inland as well as sea waters worldover. Aquaculture has come up as a very potent industry.

The kinetic energy of gravitational flow of water is tapped and transformed into electricity, what is called hydropower. This concept of electricity generation is one of the oldest and cheapest tricks largely used worldover. Morever, it leaves no pollution burden on nature as in case of thermal or nuclear power generaton.

Hardly any industry can do without water. It has multifarious application in industrial processes *viz.*, a raw material, solvent, chemical reactant, coolent, and cleaning agent.

'Water, water everywhere not any drop to drink' is the situation not only found amidst sea but also at places where industrial and domestic refuse make the benevolent water sources so severely contaminated with pathogens and or toxins that they are left unpotable. Being the best dilution medium,it is frequently used as the cheapest and convenient way for liquid and solid waste disposal. Further it was practically impossible to maintain sanitary conditions without water for it is needed in all common cleaning and washing practices.

Water at many places serves as a very convenient means of navigation. Besides, it adds to the aesthetic value of the landscape, and provides opportunities for recreation and sport.

Water being unevenly distributed over the land, influences interstate and international relations and thus is the cause of what is known as hydro-politics.

It can, therefore, be said that while water is an essential for life it is significantly linked with social, economic, political, and ecological intricacies.

Problems and Management of Water

Management of water implies making the best use available water resources for human benefit while not only preventing and controlling its depletion and degradation but also developing it in view of the present and future needs. Water, like forest, is a multipurpose resource and it is important to see that its various uses should not conflict with each other and it can be enjoyed in its totality by man and others. Thus, its right allocation, and quantitative and qualitative conservation are the primary tasks before water managers.

Floods, droughts, improper use, pollution, and disease transmission are the important problems related to water. Because usable inland water resources constitute a very minute fraction of the world's total water and because man largely depends upon this minute fraction of water for his various activities, it is important to make efficient use of available water and at the same time prevent abuse of it. This has to be considered not only in giant irrigation projects or industrial and municipal processes but also in everyday life. Taps running unpurposefully, leaking water supply pipes, using filtered water for gardens and building construction are a few instances of inefficient and misuse of water in daily life.

Water being the best and convenient dilution medium is the worst affected resource by pollution. To a common man pollution of water means only the drainage of industrial effluents or domestic sewage in natural waters. It may astonish a lay reader that even pure distilled water when released in sea may act as a pollutant since it alters the normal salinity of sea water which adversely affects the marine life. Besides chemical alterations, changes in physical condition of water also have grave effects on the biota of system as during thermal pollution caused by the discharge of heated effluents in a body of water. Pathogens, chiefly entering natural waters through municipal wastes rich in excreta and other organic filth, also render the water unsuitable for human use.

Infact, it is not only the man who has to use the water resources but these are equally important to land animals and more to those fish and other aquatic animals who live in it. Thus, pollution of a body of water does prove detrimental not only to the human society but also to regional ecology and the ecology of very aquatic ecosystem. It is, therefore, essential that the water should not be treated as a simple respository of waste disposal, and if inevitable, the wastes only after proper treatment should be released in it under strict ecological consideration.

The qualitative degradation of water is also in a way quantitative depletion usable resource. Therefore, recycling of waste water after due treatment would relieve the water scarcity to a great extent, especially in regions marred by water shortage.

Future Needs and Alternate Sources of Water

The demand for usable water, like any other resource, is increasing fast with the continuous increase in world population and continuous increase in per capital demand linked with more of sewage disposal and transport, prolific use of water-using gadgets (like air cooling and conditioning devices, washing machines etc.), and increased recreational activities. According to an estimate, the daily per capital demand for water of an American was 600 gallons in 1900 which increased to 1,500 gallons in 1960 and is expected to be increased upto 2,700 gallons by 2000. The fact is no different in developing world. Thus, the question is 'where will the water needed for fulfilling increasing demands come from ?' the answer is not so simple and can only be found through long term planning. Possible means of meeting out the impending water deficit problem include :

(i) Reclamation of sewage and waste water;

(ii) Development of ground water sources and surface storages;

(iii) Long range forecasting of rain;

(iv) Rain making;

(v) Transfer of surplus water; and

(vi) Desalination of sea water.

Agriculture

The practice through which specific plant species are cared and managed so as to obtain maximum yield of consumable parts of the plants is called agriculture. An agricultural field is, in fact, a cropland ecosystem in which ecological principles involving limiting factors and productivity provide great applicability.

The demand for food production, linked with tremendous population growth, has been increasing fast world over. This has been possible to feed increasing number of mouths through increased agricultural production, what is called 'green revolution'. It has resulted chiefly from the industrialization of agriculture, which involves highly domestic plant species, sophisticated chemicals, and large fuel energy subsidies. According to an estimate, to double the crop yield through industrialized agriculture, a ten fold increase in fertilizers, pesticides, and hoesepower is required.

The green revolution through industrialized agriculture (fuel powered) as compared to conventional agriculture (man-and-domesticated-animal powered) has following backlashes :

(i) Such an agriculture involving huge amounts of fuel, and chemicals in the form of fertilizers, pesticides and weedicides, is the chief cause of environmental pollution.

(ii) It may lead to social backlashes when a quick shift from small farms to large industrialized farms takes place. In this process smalls farmers and agriculture workers are driven off the land to cities where they may not get work, food, and housing – a situation of despair and social breakdown.

In retrospect, strong measures should be taken to control the population growth, and to improve the conventional agricultural practices with application of current agricultural research under due social and environmental consideration. For a greater and sustained agricultural production following considerations are important :

(i) Available land acreage should be properly and judiciously utilized.

(ii) Soil fertility should be increased through wise use of fertilizers and organic manures.

(iii) High yielding and disease resistant plant varieties should be introduced.

(iv) Mixed cropping should be practicised wherever possible.

(v) Integrated and balanced use of available water sources, surface and ground water, should be made.

(vi) Weeds and pests should be efficiently controlled. Integrated pest control practices should be preferred over total reliance on chemical pesticides.

(vii) Soil erosion and loss of nutrients should be prevented through maintaining vegetational cover throughout the year.

(viii) Exotic varieties should only after due consideration of long term impacts be introduced.

Forestry

A forest is a natural ecosystem having multispecies and multiaged trees as dominant community. Forests cover about one-third of the earth's land surface of which about 50 per cent is occupied by tropical forest. Of the total geographical area of India, 22.74 per cent is forest as against a minimum of 33 per cent forest cover prescribed under National Forest Policy.

Importance of Forests

Functions of a forest may broadly be classified into following three categories :

(i) Protective functions : These include the protective role of forests against soil erosion, droughts, floods, intense radiation etc.

(ii) Productive functions : Forests are the source of wood and many other products like gums, resins, fibres, medicines, katha, honey, bidiwrappers, pulp, paper etc.

(iii) Accessory functions : These include the role of forests in recreation, aesthetics, and as habitat of diverge wild-life etc.

Thus, it can be said that forests are important in two ways – ecologically and economically.

Ecological significance : Environmentally, the forests are very important. They help in balancing oxygen and carbon dioxide level in atmosphere, regulate earth's temperature regime and hydrologic cycle. Forests increase local precipitation and water holding capacity of soil, thus preventing drought situation. Vegetational cover provided by forest impedes the velocity of runoff on soil surface, checks soil erosion, silting, and landslides, thus reducing the danger of flood. The litter derived from fallen leaves maintains fertility of soil by returning the nutrients. Forests also act as refuge of wild animals and provide protection to them against strong, cold or hot and dry winds, solar radiation, rain and enemies.

Economic significance : Woods, which has various applications in domestic and industrial processes, is the chief product of forest. Wood, when used as fuel, has certain advantages over coal as its sulphur and ash contents are very low but, at the same time, excessive use of fuel wood means pressure on forests which have many other functions to serve. Wood may also be converted into solid (coke), liquid, and gaseous fuels. Timber is an important material in building construction and day-to-day uses. Forest provides raw material for various wood-based industries, *viz.*, pulp and paper, composite wood, rayon and other manmade fibres, sport goods, furniture, boat building, matches etc. Miscellaneous products, like bamboos, resins, gums, some oils, fibres, medicines, katha, lack, shellac etc., are also obtained from forests.

Depletion of Forests

Increased urbanization, industrialization, and mining has entailed indiscriminate felling of trees and denudation of forests. Nobody will, however, ever learn that who set the first assault on which tree in the history, but this practice of logging down trees for human use has been going on for centuries unabated. Colonists used logs for firewood and making huts. They also made handles for their tools, furniture, and fence posts etc., from it. For them forest had only a present and a past - no future; thus they resorted to ruthless and accelerated exploitation of available forest -- a story that still continues with more serious backlashes earth has ever witnessed. India alone is losing more than 1 5 million hectare of forest cover each year and according to Tiwari (1982) 22 million hectare of forests have been destroyed during the last three decades due to over exploitation and misuse and conversion to agricultural fields. The depleted forest wealth would simply deprive the man of economic and environmental values offered by forest, as discussed earlier.

Management of Forests

In forestry, the yield is chiefly obtained from the accumulated growth of the past, so this industry has to depend upon the annual growth. Forests are managed to obtain a sustained yield which is obviously possible through sustained growth. Following considerations are important in this direction.

(i) A forest should have in correct proportion the trees of different ages and classes so that continuous removal of exploitable produce would not affect the future yield.

(ii) Each tree crop should put on normal growth increments.

(iii) No obstacle should exist to the continuous harvest of exploitable stems.

(vi) Unrestrained felling should be prevented.

For all such planning and decision making, sound ecological knowledge is desirable that include knowledge about climatic conditions, composition of biotic community, successional trend, life cycle of important tree species, methods of arresting growth of undesirable trees and promoting growth of highyield tree species, and biotic interrelationships (such as weeds, competitions, pests, parasites and diseases etc.).

Of various tactics employed in forest management, some deserving special considerations are as follows :

(i) Restrained felling,

(ii) Block and selective cutting,

(iii) Reforestation,

(vi) Control of disasters, and

(v) Recycling of forest products.

Rangeland

A range is the grassland that provides natural pasture for grazing animals. Grasslands occur where rainfall is too low to support the forest community but is higher than that which results in desert community. Generally, in grassland precipitation ranges between 25 and 75 cm. However, depending upon edaphic factors, such as a high water table, grasslands may also occur in regions of forest climates. Soil of a grassland is characterized by having high humus content.

Over 6,000 species of grasses, adapted for a great variety of climatic regimes, cover about 20 per cent of the earth's surface. In size, they vary from as small plants as about 1 cm to over 35 cm tall bamboos.

Importance of Rangeland

Man has been cultivating various grass species as his food and as fodder for his domesticated livestock since time immemorial. History of human civilization is the witness of dependence of man on the seeds of various grasses as his own food, for instance, Indo-Chinese, American and Mediterranean cultures were based on rice, maize, and wheat respectively; while forage was used as fodder for domesticated livestock.

A range, if not serving as a direct source of food for man, still serves the mankind in a number of indirect ways. Solar energy transformed into food energy by the green cover of the range is available to primary consumers (grazers - domesticated or wild) as food, wherein a part of it is converted into meat. Of considerable importance to man are these herbivores and their products, such as meat, milk, skins, hides, horns, and wool etc. It can thus be said that at the root of all biological resources that man utilizes is the forage.

Depletion and Degradation of Rangeland

Man would not destroy an artifical pasture development by input of his hard labour and money than a natural pasture. Thus, like all natural resources, the rangeland, inflicted with human stress, is depleted and degraded. Cause of such stress, may be ignorance, greed or incompetence of man. Phenomenal growth of cattle industry, on one hand, is responsible for ruthless over grazing, while on the other hand, wide expansion of urbanization, industrialization, and agriculture has resulted in encroachment of rangelands, compelling ranchers to overstock the available limited rangeland. Often, out of greed too, a rancher allows a number of head to graze beyond the carrying capacity of the range, being ignorant to the fact that too many sick and weak head would not fetch him as much value as a few head in prime condition of health after grazing on enough good forage.

Community or ecological indicators provide the most practical approach of determining whether or not a range is being utilized properly (Humphrey, 1962). Range managers use the term "decreasers" for palatable species vulnerable to grazing; their disappearance is an indicator of grazing stress. When such indicator are not heeded, unpalatable "weeds" invade and flourish. Thus, if overgrazing continues, "decreasers" keep on disappearing replaced by "weeds". Once established, if may become hard to eradicate the weeds and not only this, such vegetation will invite a variety of pest animals like rodents and insects. This makes clear the long-term consequences of overgrazing in a range.

There is increasing need to restore the depleted and degraded rangeland not only out of economic (livestock production) but also ecological consideration since rangeland depletion would also cause problems like soil erosion, spread of pests etc.

Range Management

Stoddard and Smith (1955) defined the range management as "the science and art of planning and directing range use so as to obtain the maximum livestock production consistent with conversion of range resources".

Efforts to manage a range land include : regulation of grazing pressure; manipulation of stock (herd) distribution; burning of range; reseeding of rangeland; selection of cattle breeds; and control of weeds, pests and predators.

Wildlife

Even on global level there appears to be a great deal of ambiguity attached to the term 'wildlife'. Webster's Dictionary defines wildlife as 'living things that are neither human nor domesticated, specially mammals, birds, and fish hunted by man'. In American usage, term 'wildlife' is used to designate the game and excludes fish, thus they have Fish and Wildlife Services at the federal level. In India too, the common understanding of the people regarding wildlife is of wild animals and shikar, more particularly mammals, birds, and reptiles. However, a report entitled 'Wildlife Conservation in India' (1970) of the Indian Board for Wildlife states wildlife as the entire native uncultivated flora and fauna of the country. The Wildlife (Protection) Act, 1972 also defines wildlife as any animal, bees, butterflies, crustacea, fish and moth, and aquatic and land vegetation which form part of any habitat.

Thus, a comprehensive definition of wildlife may be given as-any or all non-cultivated and non-domesticated life including both, plant and animals.

Importance of Wildlife

The wildlife resources provide aesthetic, recreational, and economic benefits and at the same time are important ecologically. Nature's millions of forms of life on this earth perform some or the other definite biological function which keeps the delicate balance of our ecosphere. In nature flora and fauna go hand-in-hand and one cannot exist without the other. Birds and insects are necessary for the cross pollination of flowers, and these with other animals also serve for dispersal and propagation of vegetation. Not only this, the health of vegetation also depends on natural browsing and grazing. Needless to mention that animals, directly or indirectly, depend on green plants for their food. Besides food, plants also provide shelter to a variety of animal species. Equally important are the relations among animal populations. If too many herbivores are killed, the predator, deprived of their natural food, will leave forest, become vermin, and prey upon domestic lovestock or even man. On the other hand, if too many carnivores are killed, the herbivores, because of release of predation pressure, will become too numerous and acquire the status of pest for agricultural crops. Thus, most important is the balance of nature of which wildlife is an integral and important part.

Abuse and Depletion of Wildlife

One often comes across the headlines in newspapers like 'Last of the blue whales', 'Save the golden langur', 'Tragedy strikes sea turtles', 'Major downfall in rhino population', 'Great Indian bustard on the road of extinction' ...and so on, which narrate the sad story of abuse and depletion of our wildlife heritage. Accord ng to an estimate, in last about 2,000 years the world has lost, by way of extinction, about 160 species of mammals and 88 species of birds.

India is endowed with a rich biological heritage and her various scriptures and epics are full of vivid and beautiful account on wildlife. Over 500 mammalian and 1,200 avian species contribute to the richness of this wealth. Unfortunately today, the Indian wildlife, as that in many other parts of the world, is threatened because of man's unprecedented intervention with nature. Nearly 134 plant species have been declared threatened as against 71,47 and 15 species of mammals, birds and reptiles respectively (Khoshoo, 1984). Great Indian bustard Choriotis nigriceps confined to the semi-arid regions of Rajasthan and Gujarat, and the white-winged wood duck Cairina scutalata inhabiting the Brahmputra swamps and the flood plains of Assam are both on the brink of extinction.

Wildlife Management

The conservation, is the practice embracing preservation, maintenance, sustainable utilization, restoration, and enhancement of the natural environment. Wildlife conservation in this sense is the management of nature's living resources if we consider wildlife as consisting both of wild flora and fauna. Since one cannot manage living resources while being indifferent to the non-living environment, the concept of wildlife management becomes the concept of management of natural ecosystems. It also overlaps 'forestry' insofar as it is based on multiple landuse concept and implies management of wild plants as well as wild animals. However, under the present dispensation wildlife management implies management of wild animals as distinct from forestry, and it implies maintenance of habitats suitable to the different species.

In America and Europe wild animals, particularly mamals, are treated as game crops, as do timber in forestry and food crops in agriculture, and hunting is developed as a viable economic activity. In America, wildlife and wildlife related recreation forms multimillion dollar industry. However, in India it is difficult to develop game as an industry because : (i) our forest areas are already overburdened with grazing pressure of domesticated livestock and because of already scarce pasture lands there is little hope of reducing grazing pressure in the forest areas to develop game; and (ii) hunting as a source of recreation is beyond the means of the majority of population.

During past decades the wildlife management implied the understanding, increasing, and controlling of some species. Sanctuaries and National Parks were mostly meant to protect a particular species or a group of species and to a limited extent for tourism and recreation. Today, however, the emphasis is shifting from the management of species to the management of ecosystem. The objectives of wildlife management, as outlined in the World Conservation Strategy, are thus as given below :

(i) To maintain essential ecological processes and life support systems in natural ecosystems, on which human survival and development depends. These essential ecological processes and life support systems can be defined to include maintenance of both the biotic and abiotic components of the ecosystem.

(ii) To preserve genetic diversity (*i.e.*, the range and variety of genetic material found in the world's wild plants and animals) on which depend the breeding programmes necessary for the protection and improvement of cultivated plants and domesticated animals, as well as much as scientific advance, technical innovation, and the security of the many industries that use living resources.

(iii) To ensure the sustainable utilization of species and ecosystems. This will include the proper utilization of wild animals for game, forest and grazing, and to achieve proper integration for economy both of rural communities as well as trade and industry.

Fortunately, people have been alerted now and the strong measures being taken in favour of restoring, maintaining, and increasing wildlife population given hope that our rich legacy of wildlife will not only be saved for future but developed further. Ways and means employed in wildlife conservation include : (i) protection of breeding stock by enacting game laws and developing wildlife refuges; (ii) artificial stocking including introduction of exotics; (iii) control of predators; (iv) habitat improvement; and (v) game farming.

Minerals and Their Characteristics

Minerals are the nonliving substances obtained from the earth through the process of mining. Formation. of minerals took place predominantly in weak zones like mountains, folded and faulted regions, lakes, troughs, and continental shelves through geological processes going on since billions of years. They hold some peculiar characteristics, as discussed below, which distinguish them from other natural resources.

(1) Exhaustibility

Minerals are non-renewable or depleting assets and once mined out the deposits get depleted and lost for ever without any chance of replenishment. Refuse or waste, however, can be consumed afterwards through improved benefication technology.

Simply this exhaustible resource cannot be cropped year after year like agricultural products. Therefore, three distinct stages can be identified in the life cycle of a mine : youth – developmental stage; maturity—established stage; and old age—declining ore reserves. If the rate of mineral consumption is increasing fast, the knowledge about more reserves and potential substitutability of many minerals is also being improved. As such these predictions may prove outlived.

(2) Localised Occurrence

Minerals are very localised in their occurrence and the total area occupied by them is far limited. Most of the mineral deposits are very sporadic in distribution, for instance, Kolar gold fields of India occupy an area of 8 km. by 1.6 km. Formations containing coal and petroleum are, in general, bigger as compared to metalliferous deposits. The coal-bearing formations of India, for example, occupy about 4 per cent of the country's total area.

(3) Unpredictable Occurrence

The quantity and quality of a mineral in any region is not easy to predict. Uncertainities prevail with respect to their occurrence, extent of depth, cost of mining (often linked with depth), and the grade.

(4) Risk Involvement

The process of mining involves large number of problems for miners. Abnormal temperature and pressure variations, rock burst, fire and gas hazards, lack of ventilation and drainage, collapse of roof and side walls are the peculiar problems encountered only in mining industry.

(5) Dependence on Other Factors

Mineral production is largely affected by factors like the changing needs of consumer industries, availability of stock, fluctuations in prices, and reuse of metallic scrap in production process. A technologically poor country possessing richer grades of ores may have to face keen competition in international market with technologically richer countries may, however, pogsessing poorer grades or ores since in later case the cost of production is considerably low due to improved mining methods.

(6) International Character

Minerals know no political boundaries. Virtually, no country within its own boundries has adequate production and reserves of all the minerals needed for its industrial development. The unequal distribution of minerals makes the international trade in its inevitable. Some countries are bestowed with monopolistic rights in respect of certain mineral deposits which result in what is known as geopolitics.

Development and Conservation of Minerals

Geologists are of general apprehension that the known reserves of minerals may not last for long and most of them will be exhausted within 100 to 200 years. New .possibilities of mineral reserves in land and beyond the land (as in the seabed are being explored which may reduce, if not totally remove, the futuristic fears of exhaustion of the resource.

Presently, ilmenite, rutile, tin ore, diamond, sulphur, coal petroleum, and calcareous and ferruginous sea-sands are largely being mined from the sea-coasts and sea-beds. In India, ilmenite, rutile, and associated zircon, monazite, and sillimanite are obtained from the beach sands of Madras and Kerala coasts. Coral limestone for industrial use like manufacture of calcium carbide, and calcareous sand for cement are being mined from the sea-bed at Tamil Nadu and Dwarka (Gujarat) respectively for quite long time. Indonesia, Thailand, and Burma recover considerable tin ore from the sea-bed, while England recovers coal and Philippines iron from the sea-bed. A number of petroliferous structure occurring off the coast are being mined for petroleum and its products at several places. The 'ferro-manganese nodules' or what commonly known as 'manganese nodules' found over the oceanic floors, generally below 4,000 m depth, are perhaps the most spectacular discovery in this field. The most prolific concentrations of such nodules with high metal values have been found in North Pacific between 139° W to 141° W and 4° N to 16° N. Although precise information about the reserves of manganese nodules is not available but it is estimated that they may run several billion tones and may prove to be potent source of metals like manganese, nickel, copper, and cobalt. It is estimated that 1 million tons of manganese nodules from sea-bed provide about 275,000 tons of manganese, 13,000 tons of nickel, 2,500 tons of cobalt, and 1,000 tons of copper besides other metals like iron, vanadium, molybdenum, silver, and lead. Further, it has now been roughly estimated that the metal resources available in manganese nodules are several times greater than those available in land, such as cobalt 5,000 times, manganese 4,000 times, nickel 1,500 times, and copper 150 times.

The conservation of mineral resources does not mean preservation of minerals under ground to be used by future generations but it implies efforts to utilize minerals in best possible way, depending upon the industrial needs and changing technological requirements, so that the life of the deposits is increased without sacrificing out own needs. From the utilization point of view, minerals are placed under two classes : expendable and nonexpendable. Expendable minerals have practically no salvage value and once consumed are lost for ever, *e.g.*, coal, petroleum, gas, and most of the non-metallic minerals. Non-expendable minerals are those which have good salvage value and can be used again and again through recycling, except those quantities that are lost due to corrosion and abrasion, *e.g.*, metallic minerals.

Efforts to conserve mineral resources include :

(i) Application of efficient methods of mining to take out every possible tonnage lying under ground.
(ii) Utilization of unmarketable ores through the innovative methods or ore dressing and metallurgical practices.
(iii) Use and reuse of scraps.
(iv) Recovering all associated elements as co-product or by-product.
(v) Economic use of minerals.
(vi) Substituting the use of rare and costly minerals with those which are abundant and cheaper.

A facility (house, hospital, port), or even in providing a service to society (medical care, public transport). Firewood, fossil fuels (coal, petroleum and its products, natural gas), solar energy, biogass, wind and hydropower, and atomic power are chief energy resources. Of these, firewood and fossil fuel are being used for supply of energy since time immemorial and are known as conventional energy sources. Wind and hydropower, solar energy, biogas, and atomic energy are the new and non-conventional sources of energy. Fossil fuels being limited in supply may be termed as exhaustible sources while others, like solar, hydel, wind and atomic energy, that would be available to man for infinite period are regarded as inexhaustible sources. Firewood and other biofuels (such as, biogas derived from biological wastes, domestic garbage etc.) are indeed renewable energy sources but exploitation of firewood has great ecological implications since we are already short of forests and whatever forests available should not be squeezed or thinned. However, use of other biofuels is known to be the best option in the light of two important facts : their steady or even increasing supply at low cost; their use in energy production reduces pollution of environment.

Because of rapid depletion and poor regeneration of the exhaustible sources of energy the society should concentrate on tapping the inexhaustible and renewable sources under strict economic and ecological consideration.

Conventional and Exhaustible Energy Sources

Firewood : No one can trace back in the history that whose are made the first assault on which tree. However, it is evident that man has been logging down the trees for various purposes, including to get firewood as an important one, since long. Because, in olden days man's knowledge and access for various natural resources was limited, he ruthlessly exploited his own surroundings to meet out his day-to-day requirements. The story still continues in various parts of the world at slower or faster pace, but often not because of the dearth of knowledge but out of vested interests or because of having no other option. Felling of trees for firewood is still a common practice in interior where it is the cheapest and, at times, the only available source of energy. This had led to thinning or denudation of woodland that need serious consideration from both ecological and economic viewpoints. Obviously, the only wayout to the problem is making available to the villager the cheaper non-conventional energy sources, such as biogas and solar energy in usable form.

Fossile fuel : Fossil fuels (coal, petroleum, natural gas) are the most extensively used sources of energy today. Like for many other natural resources, increase in population and per capital energy demand coupled with the industrialization at a faster pace, are the factors responsible for depletion of fossil fuel world over. According to a report the earth's original stocks of coal and petroleum were about 6,000 billion tons and 1,250 billion barrels respectively, of which by the year 1960, 100 billion tons of coal was mined and 100 billion barrels petroleum extracted. The story of natural gas is no different which has acquired the status of irreplaceable commodity of domestic use in modern times. Petroleum and natural gas today fulfill 60 per cent of the world's total energy requirements. It is evident that if the exploitation and consumption of these resources continues at the present incredible rate, there supply may last only for a few more decades.

Nonconventional and Inexhaustible Energy Sources

The rapid depletion of conventional energy sources has promoted governments and people to concentrate on finding and tapping some non-conventional energy sources that may last for long. In fact, inexhaustible energy sources, like solar, hydel, tidal, wind and atomic power, can only bring hope for the sustained socio-economic development of humanity for years to come when exhaustible sources will become either inaccessible or too costly to afford.

Solar Energy. Research and application in the field of solar energy have opened new vistas in the direction of fulfilling world's future energy requirements. It has especially drawn attention for its practically free steady supply and pollution-free use. This energy may be transformed into heat using reflectors and condensors, or into electric energy using solar photovoltaics. The resource can especially be tapped more effectively in the regions where there are long sun-shine hours.

Solar heaters and cookers are becoming popular these days in industry and household. Passive solar architectures provide comfortable housing in extreme cold and hot climates. This engineering has also been successfully applied in warehouses and cold storages. A multipurpose solar dryer-cum-warehouse is under construction at Ganaur (Haryana, India) where chillies, potato-chips and other foodstuffs could be dried and stored. Solar energy has also been effectively used in desolination of water. College of Engineering, Pune has designed a solar furnance that can generate heat upto 2,000°C which can be used in various metallurgical and other purposes.

Department of Nonconventional Energy Sources of Govt. of India has prepared a perspective plan that envisages generation of energy through non-conventional sources, chiefly solar, to the extent of 250 million tons of coal replacement per annum including 15,000 MW of electricity by the year 2001. Indian's first solar power station, soon to be operated near New Delhi, would be a great breakthrough in this direction. Rural electrification through solar photovoltaics is already in operation at some places in the country. Moreover, photovoltaic irrigation pump sets have also been installed at some places.

Wind power : The power of wind since olden days is being used in running mills, irrigation of fields, and carrying out other farm activities, but its real potential is not so far being realized. According to inventories of Department of Nonconventional Energy Source of Government of India 20,000 MW electricity can be generated from the wind alone in the country. Wind power plants are set up in various parts of the country but they still account for the total capacity of 3.3 MW only. Obviously, this potent resource is still to be tapped optimally.

Hydropower : Use of gravitational flow of water in generation or electricity, what commonly known as hydropower or hydel power, is quite common in various parts of the world. In India, first hydropower station, having a potential of 3000 KW, was construted in 1897 at Darjeeling. Later on, many stations with far high potentiality were set up. The country's economically exploitable hydropower resources have been estimated at 41 million KW (or 41,000 MW) of which only 18 per cent of the potential was utilized by 1979. The Brahmaputra basin has the highest hydropower potential and nearly 30 per cent of the country's total; next to it rank the Indus, Godavari, and Ganga basins respectively.

In world, a total of $4.910 \cdot 10^9$ KW electric power is generated of which hydropower forms a fourth In highly industrialized countries like USA, USSR, Canada, Japan, France, Sweden, and Norway hydropower is nearly full developed whereas in the rest of the world it is yet to be developed fully.

Hydel power is considered to be the cheapest source of electricity with the cost of generation ranging from 2 to 3 paise per unit, compared to a minimum of 5 paise per unit of electricity generated by coalbased thermal power plants. Although the initial cost of hydel project is generally 25 per cent higher but owing to lower recurring costs, the cost of generation of hydropower is the lowest. Apart from such economic considerations, development of such projects is beneficial for irrigation and other purposes too.

Tidal power of water in coastal areas of India is, however, not so far being realized. According to an estimate a tidal power potential of 9,000 MW has been identified, of which 7,000 to 8,000 MW is in the Gulf of Cambay, 800 to 1,000 MW in the Gulf of Kutch, and rest in Sunderbans.

Atomic power (Nuclear power) : As already delineated, the fossil fuels being exhaustible, such sources of energy are not to last for long with the current trend of consumption. Atomic power appears to be the only hope for large scale energy requirements when fossil fuels are exhausted. The reserves of atomic fuels are said to be over 10 times greater than fossil fuels and the advantage of former is that their very small amounts would produce enormous amount of energy. A ton of Uranium^{235}, for instance, would provide as much energy as 3 million tons of coal of 12 million barrels of oil. Atomic energy has its application not only in generation of electricity but has successfully been used in marine vessels, spaceships, and even in chemical and food-processing industry.

Important constraints in atomic energy generation are :

(i) Cost of construction and maintenance of plants; and

(ii) Disposal of radioactive wastes. Cost of construction of an atomic power plant and its maintenance is very high as compared to any other electric power plant. For running one atomic power unit of 235 MW capacity, heavy water of a worth Rs. 10 crore is required per year. Further, as a consequence of frequent failures, the running cost of such plants come to be very high. Moreover, after the working capacity of an atomic power plant is over (in about 25 years), these have to be destroyed so as to avoid radiation spillage. Surprisingly, the cost and time for destruction of these plants is almost equal to that for construction. Disposal of radioactive wastes is such a problem that still has no satisfactory solution for long term.

Bioenergy (Energy from biological matter) : Organic wastes provide an important renewable source of energy. Biogas generated from the decomposition of organic wastes, like manure and domestic wastes, is becoming increasingly popular as a source of heat and light in many parts. It is considered to be advantageous in view of its relatively cheaper supply, and use of organic wastes in its generation reduces the othewise impending threat of pollution due to their release in environment. As such, it serves two purposes; fuel production and waste disposal. It has more of practical applicability in villages where organic waste, in the form of cattle dung, is available in plenty. Biogas so generated is thus economic and convenient to use as compared to conventional practice of burning of dry dungcakes. Further, the dung-cakes serve only as a source of heat for cooking and allied purposes while biogas has multifarious applications. The slurry so produced in a biogas plant can well be used as manure for agricultural fields.

Under the National Biogas Development Programme, 8.2 lac biogas plants were set up by the year 1985-86 in different parts of the country. A large-scale urban waste recycling plant at Okhla (New Delhi) generates about 0.6 million cubic feet gas per day having heat value of 700-800 BTU per cubic foot. The gas is being supplied to about 800 households.

Conservation of Energy

The need for conservation of energy becomes essential in

view of the following simple facts :

(i) Most of the activities of present day large and growing population require massive energy inputs, thus per capital and total demands for energy are fast increasing.

(ii) The fossil fuels, we are not endowed with which much, are not to last for long at the present rate of consumption. At the same time, much talked about possible alternate sources of energy are, in fact, still in tentative state of development so far their practical applicability is concerned.

(iii) There are economic constraints for many countries even if they want to develop conventional sources of energy, what to talk about sources like nuclear energy.

(iv) Unplanned exploitation and use of energy result in widespread environmental damage, the cost of which is to be borne by humanity.

Conservation of energy means not simply the curtailment of energy consumption but to curtail unnecessary or wasteful consumption of energy and make thrifty use of energy producing resources to ensure their steady supply for generations.

10

Endangered Species of India

ENDANGERED FLORA OF INDIA

Introduction

The increase of human population in the last few decades demanding development in various spheres has resulted directly or indirectly in the sudden and often far-reaching disturbances in natural ecosystems. The growth of large urban areas, construction activities such as dams, buildings and roads, encroachment on vast areas of forest lands for extension of arable expanses and mining operations are examples of direct onslaughts on nature which have steadily depleted natural resources. Besides, the pollution resulting from urbanization, industrialization, excessive use of chemical fertilizers, and pesticides and also of the several giant aircrafts have in no small way contributed to eco-imbalances in soil water and air. These imbalances have posed problems for the very survival of wild life. Besides these, in order to satisfy his ego and establish supremacy over other life by collecting hunting, trapping and poaching man has driven certain species on the verge of extinction.

Extinction has been the destiny of a great number of plant species including several unique and irreplaceable varieties. Some of these have disappeared from the earth in nature's own process of evolutionary change. But for many others, extinction has been caused by man because of his inadvertence as well as ignorance about their economic potential and ecological functions. So far only a small fraction of the diverse populations of plants is known to us and a large majority still remains unidentified. Even in respect of familiar species in many cases, little more is known than their appearance and location.

Through indiscriminate exploitation, destruction of habitats, spread of harmful chemicals and introduction of alien species, a number of plants have already disappeared while others await a similar fate, at times, even without our being aware of their existence. Thus human interference has been, more often than not, responsible for depletion of plant resources and consequent decline of genetic diversity. The according to an estimate of the Threatened Plants Committee of the International Union for Conservation of Nature and Natural Resources (IUCN) about 10% (20,000 to 30,000) of the world's flowering plants are reported to be "dangerously rare", or "under threat". Norman Myers, noted conservationist, concludes that at least one species is disappearing each day in the tropical forests alone and in a few more years there may well be a species lost each hour. The seriousness of the situation can be understood from the observation of Peter H Raven that a "disappearing plant can take with it 10 to 30 dependent species, such as insects, higher animals and even other plants".

Realizing the ecological importance and economic utility of plant resources, there is growing concern throughout the world about the need for conservation. As pointed out by Otto Frankel, the well known Australian geneticist, it is now time for determined efforts "to keep evolutionary options open so far as we can", and to desit from eroding the richness of genetic reserves. During the recent past such concern has fouud expression in the establishment of an International Board for Plant Genetic Resources, and the UNESCO launched the Man and the Biosphere (MAB) programme.

General Background to the Problem of Threat to Plant Species

The Forest administration in India is more than 100 years old. Efforts to identify plant species as a part of wildlife and to recognise the importance for providing special protection to endangered plant species have so far been negligible. While the rapid disappearance of some beautiful animals stimulated efforts to protect fauna, this has not been the case with flora. The word 'Wildlife' had till recently been considered synonymous with animal life and consequently, conservations and naturalists have their attention only to conservation of animal species.

It was only in the year 1968 at an International Conference (UNESCO 1968) that the problem of conservation of floappreciated and several recommendations were made urging the International Biological Programme (IBP), the International Union for Conservation of Nature and Natural Resources (IUCN) and various international and national organisations to initiate studies in to the problems involved, particularly the problem of protecting and preserving wild fauna and flora in their natural habitat/ecosystems by establishing nature reserves.

Later at the 10th General Meeting of the IUCN, the Survival Service Commission reviewed the status of endangered species of plants and their habitats. The recent promulgations of the United States Endangered Species Act (1973), the UK. Wild Creatures and Wild Plants Act (1975), development of international conventions on conservations (Wetlands Convention, World Heritage Convention, Endangered Species Convention and South Pacific Convention) and the setting up of Biological Records Centre of the Nature Conservancy, U K. and threatened plants, orchids, cycads treewide awakening of the need for the conservations preservation and protection of floras of the world.

According to the Volume 5 of Red Data Book on angiosperms started by the Survival Service Commission of the IUCN in the year 1970 (Melville, 1970), it is estimated that out of the total of three lakhs (0 3 million) species of plants in the world, over 20,000 were in the category of either endangered or vulnerable and threatened with extinction by the year 2000 A.D.

Rare Species of Plants, their Categories and Definitions

The following categories of rare plants have been recognised by the IUCN, mainly based on (i) present distribution, (ii) decline in number of population in course of time, (iii) abundance and quality of natural habitats, and (iv) biology and potential value of the species. These are defined as follows for purposes of conservation.

Endangered (E)

Taxa in danger of extinction and whose survival is unlikely if the causal factors continue operating.

Included are taxa whose numbers have been reduced to a critical level or whose habitats have been so drastically reduced that they are deemed to be in immediate danger of extinction..

Vulnerable (V)

Taxa believed likely to move into the endangered category in the near future if the causal factors continue operating.

Included are taxa, of which most or all the populations are decreasing because of over-exploitation, extensive destruction of habitat or other environmental, disturbance; taxa with populations that have been seriously depleted and whose ultimate security is not yet assured; and taxa with populations that are still abundant but are under threat from serious adverse factors throughout their range.

Rare (R)

Taxa with small world populations that are not at present endangered or vulnerable, but are at risk.

These taxes are usually localised within restricted geographical areas or habitats or thinly scattered over a more extensive range.

Threatened (T)

Threatened is used in the conservation context for species which are in one of the categories Endangered, Vulnerable and Rare. Some species are marked as threatened where it is known that they are Endangered, Vulnerable or Rare, but there is not enough information to say which of these three categories is appropriate.

Out of Danger (O)

Taxa formerly included in one of the above categories, but which are now considered relatively secure because effective conservation measures have been taken or the previous threat to their survival has been removed.

Indeterminate (I)

Taxa that are suspected of belonging to one of the first three categories, but for which insufficient information is currently available.

Main Causes of Threat to Species

It is well known now that several plant species have become extinct due to certain natural phenomena, such as land upheavals, volcanic eruptions, glaciation, protracted periods of rain or draught, spreading of desert lands, forest fires, and eutrophication in the geological past. While such natural processes in the past had no doubt led to the extinction of flora, the resulting new environmental conditions had also resulted in the evolution and speciation of new flora and migrasion of floral elements.

But, in recent times man with his anthrpopogenic associates and other factors or practices such as fire or 'slash and burn' for shifting cultivation (also called as 'jhum' or 'podu' cultivation in India), grazing by cattle and by several other mechanical means, has suddenly accelerated disastrous condition in natural ecosystems.

Besides, commercial exploitation of entire plants, roots, rhizomes, tubers bulbs seeds and fruits has been the prime cause of depletion of more important wild economic, plants throughout the world for lucrative financial gains, in the trade which flourishes both by legal and illegal means. Rauvolfia serpentina, Coptis teeta, Dioscorea spp, and Podophylum; hexandrum serve as good examples.

Area Vulnerable to Rarity and Extinction of Species in India

In India, the maximum threat to plant species lies in areas which are subjected to considerable development activity *e.g.* forests which lie in close proximity to villages and agricultural lands. Such areas occur in the entire length and breath of the Himalayas, the Western and Eastern Ghats of peninsular India, the Vindhya and Satpura Hill ranges, and the densely forested hilly region in Assam, Nagaland, Meghalaya, Manipur and Mizoram in Eastern India. These regions incidentally abound in many tourist spots and pilgrimage centres, which attract large number of people who cause damage to the flora.

Besides, endemics which are confined to narrow, restricted areas or specialized ecological niches, also in some respects fall in the category of threatened plants. Some phytogeographers believe that endemic species or genera are the survivors of the one time widely distributed groups which are now in the course of gradual extinction. Other maintain that they are newly evolved and recent forms still in the process of extending their geographical range of distribution. In such cases, due to their rarity they are to be considered as threatened plants, because any disturbance or imbalance in their narrowly confined ecosystems/habitats could mean extermination of the species.

Insolation, as in the case of island floras, and natural barriers such as mountain ranges are considered main factors resulting in endemism.

India has a very rich element of endemics in its flora. According to Chatterjee (1940) Indian subcontinent has about

61.5 per cent to endemic flora with about 7,000 endemic species and 134 endemic genera. Of these the Himalayas and the Khasi Hills account for about 3,000 and Deccan peninsula for about 2,000 endemic species. These figures speak eloquently of the great need for protecting the endemics. This can be dome effectively by carefully analyzing the floristic composition of the various phytogeographical units of Indian flora and by selecting suitable natural forests in these units and preservlng them as 'Biosphere Reserves'.

Sapria himalayana, Uvaria lurida, Alcimandra cathcartii, Magnolia gustavii, M. pealiana. Pachylarnax pleiocarpa, Nepenthes khasiana, Dicentra royler; several species of Primula and Rhododendron and the Lady's Slipper orchids. Paphiopedilum insigne. P. hirsutissimum, P. fairienum, P spicerianum and P. venustum, are some of the noteworthy endemics of the Himalayas and Khasi Hills. Antiaris toxicaria, Campanula cytinoides, Pedicularis perrotter, and the some species of the Podostemaceae are endemic to Western Ghats or the Nilgiri Hills in South India.

Himalaya and Eastern India

1. Abies delavay. Franchet (Pinaceae) Tree A Chinense Fir recorded from Arunachal Pradesh, rare occurrence.

2. Acanthephippum Sylthetense Lindi (Orchidaceae) Herb one A terrestrial orchid distributed in Garampani sanctuary (Assam Kashi, Orissa, Abor Hills and Sylhet (Bangladesh) rare occurrence.

3. Aconitum deinorrhizum Stapf (Ranunculaceae) Herb One of the important Indian Aconites of medicinal value originally discovered from Bashahr Himalaya are recently reported from Bhadrawah, J&K State rare occurrence, medicinal, exploited for made.

4. Adinandra griffithu Dyer (Theaceae) Tree Distributed in Khasi Hills, Meghalaya rare occurrence.

5 Aglaia perviridis Hiern (Meliaceae) Tree Distributed in Khasi Hills, Meghalaya, rare occurrence.

6. Amblyanthus multiflorus Mez (Myrsinaceae) Shurb Rarely reported from Assam.

7. Anacolosa ilicoides Mast (Olacaceae) Tree Distributed in Khasi Hills, Meghalaya, rare occurrence.

8. Anoectochilus sikkimensis King & Plant (orchidaceae) Terrestrial herb, the 'Jewel Orchid'. Distributed in Sikkim Himalaya, recently reported from Garampani sanctuary, Assam, rare occurrence.

9. Angopteris evecta Hoffm (Angiopteridaceae) Terrestrial fern. Widely distributed in India, Sri Lanka and sepals Asia, sporadic in occurrence.

10. Aphyllorchis montana Reichb (Orchidaceae) Herb A Terrestrial orchid, petals straw-coloured purple at back and tips. Distributed in Sikkim and Khasi Hills, rare occurrence.

11. Arachnanthe cathcartii (Lindi) Benth (Esmeralda cathcartii (Lindi) Reichb (Orchidaceae).

Epiphytic Attractive with large brownish-red flowers. Distributed in East Nepal, Sikkim, Bhutan and Arunachal Pradesh, rare occurrence.

12. Arachnanthe clarker (Reichb f f) Rolfe (Esmeralda clarker Reichb f) (Orchidaceae) Epiphytic Robust orchid with large, attractive, yellowish-brown flowers. Occurs only in Sikkim and Bhutan, rare occurrence.

13. Artemisia amygdalina Decne (Asteraceae) Herb. A botanically interesting species with leaves like those of a willow (J.D. Hooker). Known only from Pir Panjal, J&K State, extremely rare in occurrence.

14. Arundina graminifolia (D. Don) Hochr (Orchidaceae) Terrestrial, perennial plant. The Bamboo orchid with highly ornamental flowers. Distributed in Sikkim, Khasi Hills and Arunachal Pradesh, rare occurrence, horticultural value, exploited in trade.

15. Astragalus strobiliferus Royle (Papilionacea) - Herb. The only member of the Tragacanth group of Himalayan Asttragali occurring in W. Himalaya: rare occurrence.

16. Atropa acuminata Royle ex Lindl. (Solanaceae)—Herb. The 'Indian Belladonna', distributed in Himachal Pradesh and Kashmir; rare occurrence; exploited as medicinal plant.

17. Balanophora dioica R. Br. (Balanophoraceae)—Herb. Clump-forming, fleshy, club-shaped parasite on tree roots in dense forests in the entire Himalayan region; of extreme rare occurrence; botanical interest (Also other species of Balanophora).

18. Botrychium virginianum Sw (Ophioglossaceae)—Perennial. A pretty fern of botanical interest. Distributed in the Himalayas (Kumaon to Arunachal Pradesh), Khasi Hills, Peninsular India and Sri Lanka: rare occurrence.

19. Brainae insignis (Hook) J. Sm. (Blechnaceae)—Terrestrial fern. Distributed only in Khasi Hills in India; very rare; also occurs in Malay Peninsula.

20. Camellia caduca C.B.Cl.ex. Brandis (Theaceae) Shurb. Distributed in Sikkim, Bhutan, Arunachal Pradesh and Khasi Hills; rare occurrence; botanical interest.

21. Catamixis baccharoides T. Thoms. (Asteraceae)—Herb. An endemic species restricted to rock cliffs in U.P. Siwaliks; rare occurrence; botanical interest.

22. Colchicum luteum, Baker (Liliaceae)—Herb. Endemic to N.W. Himalaya; rare occurrence; exploited in trade for its valuable alkaloid Colchicine.

23. Captis teeta wall (Ranunculaceae)—Herb. Distributed in Mishmi Hills, Arunachal Pradesh, rare occurrence; heavily exploited in trade for its medicinal rhizomes—'Mishmi Teeta' now under cultivation.

24. Cyathea gigantea (Wall ex Hook). Holtt, (Cyatheaceae)—The giant tree fern of botanical interest. Widely distributed in hilly regions of eastern and peninsular India, but nowhere abundant; exploited for starch by the local people; trunk used for growing epiphytic orchids.

25. Cymbidium macrorhizon Lindl, (Orchidaceae) The only terrestrial leafless Indian Cymbidium. Distributed in

N.W. Himalaya, Sikkim, Khasi and Naga Hills; rare occurrence; horticultural importance.

26. Cypripedium cordigerum D. Don (Orchidaceae)—Terrestrial herb, highly ornamental "Lady's Slipper Orchid", once reported abundant at Gulmarg (Coverntry's Wild flowers of Kashir, 1923), but has been sighted only once in the course of five visits (K.S. Sahni). Also scarce in Western Himalaya and Tehri-Garhwal region; extremely rare; threatened due to trade.

27. Cypripedium elegans Reichb, f. (Orchidaceae) Terrestrial herb; flowers brownish, attractive. One of the ornamental "Lady's Slipper Orchids". Distributed in lachen Valley, Sikkim and eastern Tibet rare occurrence, threatened due to trade.

28. Cypripedium himalacicum Rolfe (Orchidaceae)-Terrestrail herb; flowers showy, reddish or purplish-red. Distributed in Western Himalayan, Sikkim and Bhutan, rare occurrence; threatened due to trade.

29. Cypripedium macranthon Sw, (Orchidaceae)—Terrestrial herb with attractive large brownish-purple flowers. Distributed from Garhwal to Sikkim; rare occurrence; threatened due to trade.

30. Dendrobium densiflorum Wall ex Lindl (Orchidaceae)—Epiphyte. A highly ornamental orchid distributed in the tropical and subtropical forests of Nepal Sikkim, Arunachal Pradesh and Khasi Hills but nowhere common; exploited in trade.

—31. Dianthus Coschemiricus Edgew, (Caryophyllaceae)—Herb apparently endemic in Kashmir, rare occurrence. (D. Chinesis L, is cultivated in India for its perfume-yielding flowers).

32. Didiciea cunninghamii King et Prain (Orchidaceae)—Terrestrial herb. An interesting monotypic orchid originally found in Sikkim and recently collected in Garhwal Himalayas after a lapse of seven decades; rare occurrence; botanical interest.

33. Dioscorea deltoidea wall, ex kunth (Discoreaceae). Hioscorcuceae - Herbaceous climber, Western Himalaya; becoming scarce; threatened due to over-exploitation from the wild for its medicinally important tubers.

34. Dipteris Wallichii (R.Br.) Moore (Dipteridaceae) m—Perennial herb. A fern of botanical interest. Distributed in Khasi and Jaintia Hills; recently discovered in Subansiri, Arunachal Pradesh; rare occurrence.

35. Dischidia benghalensis Colebr (Asclepiadaceae) - An interesting climber with fleshy leaves. Eastern Nepal, Sikkim, Assam rare occurrence; botanical interest.

36. Dischidia rafflesiana Wall (Asclepiadaceae)—A Climber with pitcher-like leaves, Assam; rare occurrence; botanical curio.

37. Drosera burmanns vahl Droseraceae) - Herb. The wellknown insectivorous 'Sundew' plant, Also the other species D peltata Sm. Both distributed from Himalayas to the east and south, scarcely distributed; botanical interest.

38. Drosera indica inn (Droseraceae) – Herb. In marshy habitats. Distributed in Deccan peninsula, upto Chota Nagpur, recently reported from Birbhum (W. Bengal; sporadic); botanical interest.

39. Elaeocorpus prunifolius (C. Muell.) Mast (Elaeocarpaceae) – Tree distributed in Khasi Hills and Manipur; rare occurrence; botanical interest.

40. Eremostachys superba Royle ex. Benth (Lamiaceae)—An attractive large herb on the slops of Dehra Dun Siwaliks. Threatened due to crosion of habitat.

41. Eria crassicaulis Hook f. (Orchidaceae) – Epiphyte. An attractive orchid with long recemes and pale purple-banded flowers. Distributed only in Kashi Hills; fare; horticultural value.

42. Galeola lindleyana Reichb. f. (Orchidaceae)—Tall, attractive saprophytic orchid with tomentose yellow flower. Sikkim, Kashi and Naga Hills. Also another species G. faiconeri

Hk. f. with deep yellow flowers, is found only in parts of Arunachal Pradesh; rare occurrence; botanical interest.

43. Gentiana kurroo Royle (Gentianaceae) – A herb with beautiful, large purple-blue flowers. Western Himalaya, rare occurrence, medicinally important; exploited in trade.

45. Hedysarum cachemirianum benth ex Baker (Papilionaceae) – Herb. An attractive plant Endemic in, Kashmir; rare occurrence; botanical interest.

45. Helminthostachys zeylanica (Linn) Hook (Helminthostachyaceae) – Terrestrial, perennial fern. Distributed in submontane parts of north-east and peninsular India and deltaic Bengal, recently collected in Pilibhit district, U.P. sporadic in distribution, botanical interest.

46. Helwingia hinalaica Hook. f. & Thoms ex C.B.C.I. : (helwingiraceae) – Shrub. An attractive plant with flowers. Eastern Himalaya and Khasi Hills; rare occurrence; of great botanical interest.

47. Iiex embelioides Hook. f. (Aquifoliacease) – Small tree. Distributed in Khasi Hills; rare.

48. Iodes hookeriana Baill (Lcacinaceae) - Woody climber with attractive penicles of yellow-red drupe's Kashi Hills (and Chittagong Hills, Bangladesh); rare occurrence botanical interest; horticultural value.

49. Lavatera kashmiriana Camb (Malvaceae) – Herb. 'The Kashmir Hollyhock' with pink flowers; endemic in Kashmir, scarce.

50. Lespedeza elegans Camb (Papilionaceae) – Endemic in Kashmir; sporadic in occurrence.

51. Loropetalum Chinense Olive (Hamamelidaceae) – Small tree; distributed in Assam and Arunachal Pradesh; scarce.

52. Magnolia griffithii Hook. f. & Thoms (Magnoliaceae) – Tree Endemic in Upper Assam; rare occurrence.

53. Magnolia gustavi king (Magnoliaceae) – Large tree. Endemic in upper Assam; rare occurrence.

54. Magnolia pterocarpa Roxb (Magnoliaceae) – Tall tree. Scarcely found in Assam and Arunachal Pradesh, Hutchinson (Evolution and Phylogeny of flowering plants, 1969) considers this plant to be perhaps the most ancient species of living flowering plants. Of considerable botanical interest.

55. Meconopsis betonicifolia Franchet (Papaveraceae) – Perennial herb. Blue poppy of the Eastern Himalaya also found in S.E., Tibet, Northern Burma and Yunnan; very rate in the wild; exploited for trade; introduced into Cultivation in key Botanic Gardens in 1924 and now growth in Europe.

56. Nardostachys grandifloro DC (Valerianaceae) – Perennial herb. The Himalayan spikenard with aromatic rhizomes (Jatamansi), Alpine Himalayas, rare occurrence; exploited in trade.

57. Nepenthes Khasiana Hook. f. (Nepenthaceae) – Climbing under-shrub. The Indian insectivorous pitcher plant. Endemic in Khasi and Garo Hills; rare; botanical curio. The populations have dwindled due to excessive collections by traders and students of botany.

58. Iax nana Wall (Olacaceae) – Tree. Distributed in Nepal, Darjeeling, recently reported from Assam; rare and sporadic in occurrence.

59. Ormosia glauca Wall (Papilionaceae) – Tree. Distributed in, Nepal, Darjeeling, recently reported from Assam; rare and sporadic in occurrence.

60. Osmunda regalis Linn. (Osmundaceae) Perennial shrub. The Royal Ferm. Scattered in Kumaon Eastern Himalayas, Khasi Hills, Pachmarhi Hills, and peninsular India, but nowhere abundant; botanical interest.

61. Paphiopedilum druryi (Bedd.) Pfitz. (Orchidaceae) – Terrestrial herb, flowers yellow-green with reddish-purple blotches and streaks. Endemic in Travancore Hills, W. Ghats; rare occurrence.

62. Paphiopedilum fairleanem (Lindl) Pfitz. (Orchidaceae) – Terrestrial herb. The "Asian Lady's Slipper Orchid", with large attractive greenish or cremy, reddist and purple-veined

flowers. Endemic in Kameng (Arunachal Pradesh and Bhutan; very rare occurrence threatened due to habitat destruction and indiscriminate collections for trade.

63. Paphiopedilum hirsutissmum (Lindl) Pfitz. (Orchidaceae)—Terrestrial herb with large violet-purple flowers of ornamental value. Endemic in Khasi Hills, becoming extremely rare due top indiscriminate collections from the wild.

64. Paphiopedilum insigne (Wall) Pfitz. (Orchidaceae)—Terrestrial herb with attractive greenish, purplespotted flowers. Sparsely distributed in E. Himalayas, Khasi Hills and Nepal; becoming every scarce due to habitat destruction and exploitation in trade.

65. Paphiopedilum spicerianum (Reichb. F.) Pfitz. (Orchidaceae) –Terrestrial Scapigerous herb; flowers with purplish sepals and yellow or olive green, red-spotted petals and brownish-crimson tinegd lip. Endemic in cachar Hills, Assam; rare occurrence.

66. Paphiopedilum venustum (Lindl) Pfitz. (Orchidaceae)—Terrestrial herb with blotched leaves; flowers sowny and ornamental with white dark-striped sepals and greenish puple warty-petals distributed in tropical Sikkim Himalayas and Meghalaya, becoming rare due to clearing of forests and over-exploitation from the wild.

67. Paphiopedilum villosum (Lindl) Pfitz. (Orchidaceae)–Terrestrial herb, flowers large, ornamental publish-brown, distributed in Mizoram (Lushal Hills) and Moulmein, Tenasserim (Burma); extremely rare and threatened due to indiscriminate collections from the wild.

68. Phyllostachys bambusoides Sieb & Zucc (Proaccae)–Shrub. A Sino-Japanese bamboo, distributed in Arunachal Pradesh; rare occurrence.

69. Picea brachytyla Pritz (Pinaceae) -Tree. A Chinese spruce with restricted distribution in Arunachal Pradesh: also occurs in N. Burma, Western Hupeh and W. Yunnan (China).

70. Platycerium wallichii Hook (Polyphyllaceae)–Epiphytic fern. Recently collected from Manipur; very rare; earlier known from Malay Peninsula, Tenasserim.

71. Pleione humilis D. Don (Orchidaceae)– Herb A pretty orchid distributed in Khasi Hills and Arunachal Pradesh, recently reported from Kumaon, W. Himalayas; rare occurrences; exploited in trade.

72. Podophyllum hexandrum Royle (Podophyllaceae)—Herb. The Himalayan many-apple widely distributed, but nowhere abundant. Heavily exploited for its medicinal value.

73. Populus gamblei Dode (Salicxceae)—Tree. 'Poplar distributed in Eastern Himalayas, rare.

74 Potameia paradoxa (Hook) Koster (Lauraceae)—Tree. A curious plant with 1-celled anthers; only once collected from Bhutan: rare occurrence; botanical interest.

75. Psilotum nudun (Linn). Beauv (Psilocaceae) - Perennial herb. A fern-ally of botanical interest. Widely distributed in India but nowhere common, botanical interest.

76. Rauvolfia serpentina (Linn). Benth ex Kruz (Apocynaceae) - Shrub. A very important medicinal plant of widespread distribution. Now scarce in the wild due to over-exploitation; cultivated, 7.

77. Renanthera imschootiana Rolfe (Orchidaceae) Epiphytic herb. A climbing orchid with showy yellow red-spotted petals and red-flushed lateral sepals. Distributed in Assam region, (also in Indo-China); rare occurrence of ornamental value.

78. Rheum nobile Hook. f. and Thoms (Polygonaceae) Perenial herb. The "rhubarb' of eastern Himalayas.(400-45:0 m); rare; exploited for its leaves used as substitute for tobacco.

89. Rhododendron arizelum Balf and Forr. (Ericaceae) - Tree with highly ornamentla rose-pink flower bunches. Discovered as new to India for Subansiri district, Arunachal Pradesh; rare occurrence; not reported from other parts of the eastern Himalayas.

80. Rhododendron dalhousiae Hook. f. (Ericaceae) - A shrub by rhododendron with attractive large flowers; occurs in Eastern Himalayas; rare occurrence; exploited for horticultural purpose.

81. Rhododendron edgeworthii Hook f. (Ericaceae) - An Epiphytic ornamental shrub, reported from Arunachal Pradesh; rarely collected; exploted for horticulture value.

82. Rhododendron nivale Hook. f. (Ericaceae) – Epiphytic herb. The smallest of rhododendrons only 5 cm. high, distributed in eastern Himalayas at an altitude of 5000 m; rarely collected; botanical interest.

83. Rhododendron nutallii Booth (Ericaceae)—Tree or epiphytic shrub; flowers large white with creamy base, sweet scented. Distributed in Bhutan and Daphla Hills, Arunachal Pradesh; rarely collected.

84. Rhododendron Santapaui Sastry et al. (Ericaceae)—Pretty epiphytic shrub with white starry flowers. Collected so far only from Subansiri district, Arunachal Pradesh; very rare; endangered due to clearing of forests.

85. Rhododendron stenaulum Balf. and W.W. Sm. (Ericaceae) – Tree. Flowers white, fragrant. Discovered as new to India from Apathani valley, Subansiri District, Arunachal Pradesh. Only one tree was seen; rare occurrence; threatened due to the clearing of forests in the Valley.

86. Rhus hookeri Sahni and Bahadur (Anacardiaceae)—Small tree. Endemic in Eastern India; rare occurrence.

87. Sapriya himalayana Griff. (Rafflesiaceae) - A root parasite of botanical interest, flowers ca 15 cm across. First collected in 1836 from Mishmi Hills, then collected after a century from Aka Hills, recently collected from Lohit district; distributed only in Arunachal Pradesh; very rare; of considerable botanical interest.

88. Saussurea bracteata Decne. (Asteraceae) · Herb. A deli cately fragrant species of western Himalayas; rare occurrence.

89. Saussurea gnaphalodes (Royle) Sch-Bip (Asteraceae)—A curious woolly herb of high altitude slopes of western Himalayas; rare occurrence.

90. Saussurea lappa C.B.C.I. (Asteraceae)—Herb. Endemic to Kashmir and Himachal Pradesh. Extensively exploited from

the wild in the past; scarce and needs protection in the wild. The 'Kuth' of commerce now under cultivation.

91. Schizaea digitata Sw. (Schizaeaceae)-- A perennial fern with discontinuous distribution in Khasi Hills. Sri Lanka and Malaysia; rare occurrence; of botanical interest.

92. Tetracentron sinense Oliv. var himalense Hara and Kanai (Tetracentraceae)—A primitive vesselless dicotyledonous plant. Distributed in eastern Nepal. Bhutan and Arunachal Pradesh; rare occurrence; of considerable botanical interest.

93. Thylacospermum rupifragrum Schrenk (Caryophyllaceae)-Herb. Montypic cushion-forming plant of alpine Himalaya; rare; of botanical interest.

94. Vanda coerulea Griff (Orchidaceae)—Epiphytic. 'Blue vanda' of Asia, most beautiful of vandas. Distributed in Khasi and Jaintia Hills and Naga Hills; recently discovered in Tirap district, Arunachal Pradesh. Excessively exploited from the wild for ornamental purpose; now very scarce in the wild.

95. Vanda pumila Hook. f. (Orchidaceae) - Epiphytic herb. Occassionally found in Sikkim and Bhutan; rare in the W. Himalaya.

96. Vanilla pilifera Holtt. (Orchidaceae)—Climbing orchid with fleshy leaves and creamy flowers. Garampani Sanctuary (Assam); very rare; also distributed in Malaya.

97. Viola Falconeri Hook. f. and Thoms (Violaceae)—A pretty herb. Endemic in Kashmir; rare occurrence.

98. Zanthoxylum scandens Bl. (Rutaceae)—Straggling shrub, distributed in Khasi Hills; rare occurrence.

Rajasthan and Gujarat

99. Commiphora wightii (Arn.) Bhandari (Burseraceae)—Small tree. Exploited in the past for its gumresin from the ark. Distributed in Rajasthan and Gujarat in India; reportedly becoming scarce due to over exploitation; presently cultivated.

100. Helichrysum cutchicum (C.B.Cl) Rao et Deshpande (Asteraceae)--herb. Endemic in Gujarat; rare occurrence.

101. Hyphaene dichotoma (Baker) Furtado (H. indica Becc.) (Arecaceae) – The Indian doum palm, curiously branching. Distributed only in W. coast regions; of botanical interest.

102. Rosa involucrata Roxb. (R Lyellii Lindl. 3 (Rosaceae) – Subscandent shrub. Widely distributed in India; reportedly becoming scarce at Mt. Abu in Rajasthan due to clearing of forests.

Gangetic Plain

103. Aldrovanda vesiculosa Linn. (Droseraceae) - Aquatic herb, probably monotypic, carnivorous plant of botanical interest. Reported from salt pans near Calcutta; recently reported from Manipur and Bangladesh. Its habitat near Calcutta (Salt Lake area) is threatened due to reclamation.

Peninsular India

104. Anemia tomentosa Sw. (Schizaeceae)—Terrestrial fern. Distributed in Nilgiri, Annamalai, Palney Hills; rare occurrence.

105. Apama barberi Gamble (Aristolochiaceae) – Shrub. Endemic in Tirunelveli Hills; rare occurrence.

106. Bentinkia condapanna Borty ex. Roxb (Arecaceae) – A reed-like palm. Endemic in Kerala and Tirunelveli Hills; of botanical interest; rare occurrence.

107. Ceropegia fantastica Sedgw (Asclepiadaceae) Slender climber, described by Sedgwick in 1921, based on a single collection from N Kanark, and known only form Holotype (Kew) until 1963, when this speeies was again collected in Goa (BSIWC); very rare.

108. Cycas beddomei Dyer (Cycadaceae)--Palm like tree. Endemic to Cuddapah Hils, S. India; rare.

109. Dioscorea wightii Hook. f. (Dioscoreaceae)—Twinning climber with tuberous roots. Endemic in Tirunelveli Hills and recently collected after 100 years from Type locality; very rare.

110. Entada pursaetha DC. (Mimosaceae) Large woody climber with gigantic pods of botanical interest. Distributed

in Himalayas. Assam and W. Peninsula : becoming rare due to large scale clearing of forests.

111. Frerea indica Dalz. (Asclepiadaceae) - undershrub, Endemic in W. Ghats, S. India, rare occurrence.

112. Gnetum ula Brongn. (Gnetaceae) – Extensive woody climbing gymnosperm of botanical interest. Chiefly distributed in W. Ghats : becoming rare due to large scale clearing of forests.

113. Hoya-wightii Hook. f. (Asclepiadaceae)– Large perennial climber, often becoming totally epiphytic Endemic to W. Peninsula becoming scarce due to clearing of forests and habitat destruction.

114. Lilium neilgherrense Wt. (Liliaceae) – The Nilgiri-lily. Endemic to the Niligiris and a few high mountain ranges in S. India; rare occurrence.

115. Lobelia nicotianaefolia Roth ex Roem. and Schult. (L. excelsa Leschen) (Lobeliaceae) – Herb, distributed in W. Ghats; rare occurrence.

116. Loeseneriella bourdillonii (Gamble) Ramam. (Hippocratea bourdillonii Gamble) (Hippocrateacea) – Climbing shrub. Endemic to S.W. India; rare occurrence.

117. Manisuris devergens (Hack) Kuntz. (Poaceae) - Tufted grass. Endemic to peninsular India; collected after 130 years from Coorg district; rare occurrence. All the 12 species of Manisuris found in India are endemic; ten of them confined to South India.

118. Piper barberi Gamble (Piperaceae) - Climbing shrub with pinnately-veined leaves; endemic in Tirunelveli Hills; botanical interest; rare occurrence.

119. Podocarpus wallichianus (Podocarpaceae) – Tree. The only native conifer of peninsular India, also occurs in E. Himalayas, Khasi Hills and Nicobar Islands; rare occurrence.

120. Pterocarpus santalinus Linn. f. (Papilionaceae) – Tree. Endemic in Cuddapah district, S. India; the Red Sanders of

Commerce, exploited much in the past for its beautiful red wavygrained wood. Export already banned.

121. Pterospermum obtusifolium Wt. (Sterculiaceae) – Tree. Distributed in Malabar and Karnataka; rare occurrence.

122. Santalum album Linn. (Santalaceae) -- Tree. Usually semi-parasitic. Occurs in peninsular India. Sandlwood of commerce, much exploited in trade, becoming rare (now cultivated).

Andaman & Nicobar Islands

123. Ailanthus kurzii Prain (Simaroubaceae) – Tall tree. Endemic in Andaman Islands; becoming rare due to clearing of forests and not collected during the recent botanical explorations in these islands.

124. Canarium manii King (Burseraceae) – Tree. Distributed in Middle Andaman Island. Parkinson reports it as not uncommon but this species has not collected during the recent botanical explorations; probably becoming rare.

125. Dipterocarpus kerrii Kink (Dipterocarpaceae) Tall tree. Occurs in South Andaman. "Seen only in the Goplakabang Valley, where a sample plot of this has been made" (Parkinson, For Fl. Andaman Island, P. 93, 1923). This species has not been collected during the recent botanical explorations in Andaman Islands; rare.

126. Hippocratea nicobarica kurz (Hippocateaceae)— Climbing shrub, distributed in tropical forests of eastern coast of Katchall Island (Nicobar group); rare occurrence; not collected during the recent botanical explorations.

127. Lagerstroemia hypoleuca Kurz (Lythraceae) – Tree. Widely distributed in Andamans. This species is one of the important timber yielding trees and has been exploited much; becoming scarce in wild.

128. Myristica andamanica Hook. f. (Nyristicaceae) - Tree. Distributed in middle Andaman, South Andaman and Porlob

Islands; becoming scarce due to clearing of forests. Not collected in the recent botanical explorations.

129. Ophioglossum pendulum Linn. (Ophioglossaceae)—Epiphytic herb with long ribbon like fronds. Recently reported from Campbell Bay, Great Nicobar Islands; rare occurrence of botanical interest.

130. Podocarpus nerifolius D. Don (Podocarpaceae)—Tree. The only confier of the Andamans, although widely distributed; exploited for its valuable timber; becoming scarce in these Islands.

131. Psilotum complanatum Sw. (Psilotaceae)—Epiphytic pendulous herb, recently reported from Great Nicobar Island; rare occurrence; of botanical interest.

132. Symplocos odoratissima (Bl.) Choisy ex Zoll (S. Chengapae Raizada and Sahni) (Symplocaceae)—A shrub or tree. Distributed in Great Nicobar Island; rare occurrence. Also occurs in Indonesia, Malaya, Singapore and Philippines.

133. Uvaria nicobarica Raizada and Sahni (Annonaceae)—Woody climber, Endemic in Great Nicobar Island; rare occurrence.

Gaps in our Knowledge Rare and Endangered Indian Plants

The main lacunae in our knowledge with regard to this subject lies in the present actual status of the species *i.e.* the exact number and location of populations of these species occurring in the wild, quantities or the numbers collected for economic exploitation or other purposes; their capacity to reproduce and propagate themselves in nature and their population's or representatives in cultivation under artificial conditions in farms or experimental gardens.

Such observations can be made by repeated visits to the field and need concerted efforts of several botanists and constant monitoring. This naturally an enormous task for a country like India of the size of a sub-continent. A recent close study of the orchid genus coelogyne serves as a good example four-species of the genus *viz.* O. albo-lutea, C, treutleri, C. brevifolia and C. assamica, are known only by their mention in literatre and types (specimens/illustrations). These could not be located even in their type localities, inspite of several recent botanical explorations and are believed to have become extinct. Excepting the types, even their preserved specimen are not available in any herbarium. Only such intensive library, field and herbarium studies can reveal the real status of our threatened taxa. Similarly the recent discovery of the Indian pitcher plant, Nepenthes khasiana in Garo Hills near Tura, a locality other than its own type locality near Jowai, also highlights the need for intensive botanical explorations and constant monitoring of the floras in the country.

Action Taken to save the Rarities of Indian Flora

On the recommendations of the Botanical Survey of India and other concerned agencies, the Government of India has imposed restrictions or a complete ban on the export of certain species. The Wildlife Wardens/Chief Conservations of forests in such State have been appointed as custodians of flora and fauna to check illegal trade practices in these species. They can be exported only when certified by the specified authority (presently the Director, Forestry & Wildlife).

India is one of the signatories to the Convention on International Trade in Endangered Species of Wild Fauna and Flora (CITES, 1973), it has been ratified so far by 67 countries of the world, for strict regulation and control of trade. According to the Convention, if a pecies is even smuggled out of a country. The Country destination will co-operate by obstructing the import of the consignment.

The National committee on Environmental Planning and Coordination (NCEPC) and the National Committee on Man and Biosphere (MAB) have also been concerned with the protection habitats which have undisturbed natural vegetations several such areas have been identified for conservation and preservation as 'Biosphere Reserves', throughout the country. It is also proposed to set up Germ Plasm Banks and Gene Sanctuaries to protect the areas rich in wild relatives of certain economic and cultivated plants.

The Botanical Survey of India with its main Botanic Garden at Sibpur and the regional experimental gardens established at different altitudinal situations to suit the acclimatizations of plants occurring in different ecological conditions in the country, have in their holdings rare, endemic and threatened plants. Several species of orchids are also under cultivation and protection in the National Orchidaria at Shillong, Yercaud and Calcutta. Of these, Vanda Coerulea (Blue Vanda), Dendrobium nobile, cymbidium grandiflorum, Renanthera and species of paphiopedilum and pleione growing in shillong deserve special mention.

The Government of Meghalaya State has taken action on the recommendations of the Botanical Survey cf India, and declared the Nepenthes khasiana habitats as 'protected areas' to preserve the plants in situ.

Future Programme of Work in the Botanical Survey of India

In order to fill the gaps in our knowledge on threatened plants the following studies are being undertaken by the Survey.

1. A review of literature and study of herbarium specimens of endangered or rare species deposited in the various herbana of the country to collect data on the distribution, abundance and frequency.

2. Fresh explorations to areas for collection of plants and data on rare/endangered species with regard to the present distribution frequency and phenological data.

3. Collection of live plants for introduction into the botanical gardens for protection and further studies on propagation.

4. Preparation of distribution maps of rare and endangered plants.

5. Preparation of an illustrated book on rare and endagered plants.
6. Preparation of an identification manual of plants banned for export to facilitate proper identification and for education public.

ENDANGERED FAUNA OF INDIA

Introduction

As many as 500 million kinds of plant, animal and micro-organism have made this planet home since life began over 3.5 billion year ago. Today, there are is only 5 million to 10 million species alive we do not know exactly how many, because there are many biologically uncharted areas such as the tropical rain forests, where some estimate that such as the tropical rain forests, where some estimate that over 90% of the living organisms remain unclassified. Thus, since life began, about 490 million species have become extinct.

A Study of the history of life from an evolutionary perspective teacher us that biological extinction is a fact of life. As species evolve, the old give way to the new. Many of the so-called extinct species are represented today by their descendants. Thus, many species did not die off as such but were simply altered during evolution.

Other species vanished completely as a result of drastic climatic changes or increasing environmental resistance created by excess predation or disease. The dinosaurs, for example, perished rather abruptly all over the world 65 million years ago, possibly as a result of global cooling.

Much more recently, at the end of the pleistocene epoch 10,000 years ago, many large mammals such as the sabertooed tiger and giant tree sloth vanished from North America. Their extinction resulted either from climatic changes at the end of the Ice Age or possibly from Siberia. As these people moved south and east, they hunted the large mammals such as the wooly mammoth wooly rhinoceros, giant deer, ground sloth, mastodons, giant beavers, sabertoothed tigers, bison, and elk. They undoubtedly also competed for the same food resources.

From the end of Pleistocene until now, the earth's climate has been fairly constant, and relatively few plants and animals have become extinct in the 1800s, 84 animals species and subspecies are known to have vanished. By 1984 the number had risen to 184, and if trends continue, many scientists believe, 170 more species will disappear by 2000, bringing to total to more than 350 if conditions turn considerably worse. More than 700 additional species might be destroyed before the end of the century.

One vertebrate species is becoming extinct every nine months, compared with a natural rate of one species every thousand years. When plants, insects, and microbes are added, the rate may be as high as one species per day. Ecologists argue that plant extinctions will have much more profound impacts on the ecosystem than the extinction of animals, because plants from the base of the food web.

Some authors assert that if the world's population continues to grow and nations continue to destroy wildlife habitat by expanding agriculture and other activities at the current pace, the extinction rate could reach 20,000 species a year. At this rate 400,000 species of plants and animal could be wiped out in 20 years. World wildlife expert Norman Myers argues that we could lose 1 million species by 2000. Although such high rates of extinction are probably exaggerated extinction will undoubtedly increase unless we make conscious efforts to curb population growth and properly manage our renewable and nonrenewable resources.

Why Save Endangered Species ?

Wildlife expert Norman Myers once wrote : "We can marvel at the colours of a butterfly the grace of giraffe, the power of an elephant, the delicate structure of a dia-tom. Every time a species goes extinct, we are irreversibly impoverished." Aesthetics in itself is a good argument for saving species. As we preserve are collection for future generations, so we should preserve this living art form that painters and sculptors only dream of re-creating.

A second reason for protecting species is that other organisms have a right to live. Can we not guarantee them the right to exist, something that we ourselves doggedly strive for ?

The third reason for preserving other organisms is a matter of ethical responsibility. As a species capable of destroying the world, some ecologists and philosophers feel, we have a moral obligation to preserve organisms with lesser powers.

The fourth reason for protection is that it makes good economic sense to do so. As Myers wrote : "From morning coffee to evening nightcap we benefit in our daily life style from the fellow species that share our One Earth home. Without knowing it, we utilize hundreds of products each day that owe their origin to wild animals and plants. Indeed our welfare is intimately tied up with the welfare of wildlife. Well may conservationists proclaim that by saving the lives of wild species, we may be saving our own.

From the biosphere we reap countless benefits : fish from the sea, medicines and other products from plant important plant and animal genes needed to improve domestic crops and livestock, something for hunters to shoot at something for anglers to hook and reel in some thing for bird-watchers to peer at and research animals that help provide valuable insights into human physiology and behaviour.

The economic benefits of these activities can be enormous. In 1983 alone trout and salmon anglers spent an estimated $5 billion on equipment, transportation food and lodging and hunters and bird watchers spent billions on tennis shoes, supplies, books gasoline food and lodging. Billions of dollars pour into companies that cut timber or extract valuable medicines from plants.

Each time we take a prescription drug, the chance are one in two that its origin was a wild plant. The commercial value of such drugs is around $15 billion per year in the United Shates and about $40 billion worldwide.

The U.S. Department of Agriculture estimates that each year genes bred into commercial crops yield over $1 billion worth of food. Similar gains can be documented for Canada, Great Britain, the Soveit Union, Australia, and other agricultural nations. About half of the increased productivity in corn over the past 50 years has resulted from "genetic transfusions" from wild relatives of corn or primitive cultivars in use only in isolated regions. Some biologists argue that the current high-yielding corn, which produces 250 bushels per hectare (650 bushels per acre,) could be increased fivefold with further genetic transfusions from tis ancestors.

New developments are on the horizon that may offer financial gains and a more healthful life for us. For example, the adhesive that barnacles use to adhere to the bottoms of ships may provide us with a substance to cement tooth fillings in place. A chemical derived from the skeletons of shrimps, crabs and lobsters (chitosanase) may serve as a preventive medicine against fungi infections. A new drug that kills viruses may be developed from a Caribbean sponge till now physicians have been virtually helpless in the battle against viruses.

The future of the biological world is one of grave economic concern. The loss of fish from acid rain and over fishing results in a loss of income to owners of fishing lodges and sporting goods stores. Ecological impoverishment will keep more of us at home, diminishing income to the recreation and tourism industry.

Plants and animals lost before they are explored for possible benefits will diminish our opportunities to fight disease and increase productivity. A dozen genetic reservoirs (so-called centers of diversity) located in the tropics and subtropics are the source of virtually all commercially value plants and animals. And they provide genetic material needed in the continual battle to improve plant and animals resistance. Loss of these centers would have a global impact on food supplies.

The final reason for preserving species is to protect ecosystem stability. All living organisms are an integral part of the biosphere and provide invaluable services. These include the control of pests, the recycling of nutrients, the replenishment of atmospheric oxygen, the maintenance of local climate, the control of groundwater levels, the regulation of atmospheric carbon dioxide and global climate and the control of flooding. It is no exaggeration to say that civilization depends on the survival of the biological world. Perhaps in the short term a species lost here and there may be of little consequence for overall ecosystem stability but in the long term the cumulative effect of such losses may someday threaten our existence.

Critics of the animal protection movement argue that too much effort is directed to the saving to endangered species. Why was there so much fuss over the tinysnail darter a fish imperiled by the Tellico Dam built by the Tennesse Valley Authority (TVA) Surely this one species can have no great significance to us although no irreparable ecosystem damage would result from its extinction, the question that arises is where to draw the line. Wildlife advocates argue that if we take the attitude that each species by itself is dispensable, bit by bit we will destroy the rich biological world we live in. Somewhere the line has to be drawn each endangered species is worth saving because it stops the momentum toward widespread destruction. In growth-oriented societies this momentum may be difficult to stop, much less slow down. Therefore each hurdle put in its way becomes an important force in saving the living creatures that make up our web of life.

The Indian Scene

At one time our wild life was as remarkable for its abundance as it is even now for its variety, with tigers haunting the

the forests in hundreds, and antelopes the open plains in thousands upon thousands. Over the centuries it has played an important role in our economy and culture-as a source of protein for the masses, as items of trade, and as an inspiration for a rich religious lore. The wild life has gradually declined over the millenia, and very rapidly since the British took over the country. At last we have reached a critical stage at which most, if not all of our wild life appears on the verge of being totally wiped out. Only bold and imaginative action in conservation and management will enable us to save this rich heritage from an irreversible loss at this critical juncture.

The Indian subcontient may be divided into the following eight regions for our purpose :

(1) Punjab and the adjoining semi-arid and desert regions
(2) The alluvial plains of Ganges and lower Brahmaputra
(3) The western peninsular tract excluding the Western Ghats
(4) The eastern peninsular tract
(5) The Western Ghats or Sahoadris
(6) Ladakh and adjacent regions
(7) The forested hill-slopes of Himalayas
(8) Assam and adjacent hill tracts.

Each of these ecological regions has been subjected to distinctive demagogic precssures, commercial exploitation, agricultural expansion and other modern development activities, which are so much a part of the present day industrial and agricultural economy.

Today, the position of wild animals is, ingeneral, alarming, and many species are greatly endangered and are fighting a losing battle for survival. Happily, a general awareness has developed among the informed public about the necessity to conserve the natural ecosystems of the country and the living wild communities of animals and plants these contain. To ensure this, certain legal steps to regulate exploitation of wildlife have been taken by the Central and state Governments. Endangered and threatened species of animals are totally protected, and barring vermin of very common species, exploitation of the rest is regulated by law under the provisions of the Wildlife (Protection) Act, 1972.

To grain a full appreciation of the endangered fauna of the country, a brief account of the ocurrence of different species of Mammals, Birds, Reptiles, Amphibians, and Insects in the country and their endangered forms, is presented below.

Mammals

About 350 species constitute the mammalian fauna of India of which 81 are greatly enangered.

The order primates are represented by 19 species, of which small and slender-bodied, and is restricted to the tropical rain forests and open wood land forset of South India, while the slow loris (Nycticebus coucang) is slightly bigger and inhabits the denser forests of north-east India. They are persecuted chiefly for their flsh, and for the supposed medicinal value of various parts of the body. Among the macaques the liontailed (Macacă silenus) is one of the world's most endangered primates, surviving in evengreen forests of the Western Ghats of South India, its total population is estimated to be about 800 only the pigtailed (M. assamensis) occur in the dense evergreen forests of the north-eastern India, while the crabeating (M. fascicularis) is restricted to a few islands in the Nicobar group. All of them are exploited for economic purposes and the first two species for their flesh and for alleged medicinal value. In the langur group, the Nilgiri langur (Presbvtis johni) is a multihabitat species occurring in addition to the Sholas, in the temperature evergreen forests above c 1700 mt. altitude in the Western Ghats. Its status is not as precarious as that of the liontailed macaque, it is hunted for its flesh or supporsed medicinal properties, the skin used for making drums, its, fur is also prized. The capped (P. pileatus), the golden (P. geei) and Phayre's leafmonkey (P. phayrei) inhabit tropical and deciduous dense forests in the north-eastern India. The only ape found in India is the Hoolock gibbon (Hylobates hoolock), which thrives in the hilly forests of north-eastern India. It is hunted by local tribes for its flesh and trapped for sale.

Of the family Ursidae, the Himalayan brown Bear (Ursus arctos) occurs in the upper reaches of the Himalaya from Kashmir east to Sikkim, occassionally coming down to lower reaches in severe winter the Malayan sun Bear (Helarctos malayanus) inhabits the hilly terrains of north-eastern India, south of the Brahmaputra, and the sloth Bear (Melursus ursinus) in the well-wooded forests all over India. They are hunted for their fur and gall bladder which is supposed to have medicinal properties, also for flesh, fat and other derivatives.

The family Mustelidae has 14 species in India, of which the Ermine (Mustelidae erminea) occurs in the western Himalaya and is hunted for its pelt; whilt the ratel (Mellivora capensis is found all over the country and is killed mainly to protect poultry from it and to some extent for its skin.

The family Viverridae is represented by 15 species, of which the Malabar Civet (Viverra megaspilla) occurs in the coastal districts of Kerala and is feared to have been nearly exterminated; the spotted linsang or tiger-civet (Prionodon pardicolor) ranges from Nepal east to Arunachal Pradesh extending further eastword in the mountain and hill forests, but is nowwhere common.

The striped hyaena (Hyaena hyaena) is common in the open country with low hills and ravines throughout India. It is persecuted for its habit of predating on livestock and for occasional childlifting.

There are 15 Indian species in the family Felidae, of which the tiger (Panthera tigris) is a multihabitat species found practically throughout the country except the desert region, even upto c. 3050 mt. height in the Himalaya. Unrestricted killing for trophies, skins for commercial purposes, flesh and fat for medicinal use, derivatives of body parts for their supposed magical properties, and sometimes through eradication of proclaimed animals as man-eater or Cattle-lifter, had declined its population to 1827. Through 'Project Tiger', its population has now increased to 2278 within a few years. The Indian lion (P. leo) is restricted to the Gir forest in Gujarat. The leopard (P. pardus) and the show leopard (P. uncia) are two other big cats. The former is widely distributed throughout the country in wooded forests, while the latter is restricted to higher reaches all along the Himalaya, from 2000 to 4000 mt. altitude. These have been largely hunted for commercial purposes or for the protection of livestock. Among the lesser cats, the desert cat (Felis silverstris), the lynx (F. lynx), and the caracal (F. caracal) are found in the arid and semi-arid tracts of the western and north-western India while the jungle cat (F. chaus) and the leopard cat (F. bengalensis) are found throughout the country in small numbers. Many species of cats are hunted for their pelt. The cheetah or hunting leopard (Acinonyx jubatus) which once occurred in the plains and the hilly regions of northern, central and peninsular India, south to Karnataka, is already extinct from India.

The dugong (Dugong dugong) among the Sirenia, occurs in the Gult of Kutch, on the coast of Malbar, Gulf of Manaar and the Andaman and Nicobar Islands. Once an important dugong 'fishery' existed in the Gulf of Manaar for its flesh, but the species are now scarce there.

Among the Perissodactyla, the Great Indian one-horned rhinoceros (Rhinoceros unicornis) as once extensively distributed in the Gangetic plains, but is now restricted in India to North Bengal and the plains of Assam. It was on the verge of extinction in the beginning of this century when 1904, only about a dozen individuals remained in Kaziranga (Assam) and fewer in North Bengal. It was in great demand for the alleged aphrodisiac properties of its horn and almost every part of the animal including flesh was in great demand. Though the most sought after part of the rhino was its horn, its urine was also credited with antiseptic properties. It became the victim of unrestricted killing and trapping. Measures for conservation adopted since 1904 have helped to rehabilitate it, and today its population is estimated to be around 1000 in the Indian territory. The Indian wild ass (Asinus fiemionus) is restricted to the saline flats of the Rann of Kutch, Its population has dwindled greatly from 5000 in 1946 to 870 in 1972. The present population is estimated to be 720. The major reasons for its decline are probably Surra disease and regular disturbance in their habitat by the saltworkers.

In the family Cervidae, out of the nine species; five are endangered. These are Kashmir stag or hangul (Cervus elaphus hanglu) which is restricted to the northern side of the Kashmir Valley and some adjacent valleys, its population is estimated to be about 100 in Dachigam near Srinagar and another 200 elsewhere is Dessu in the Bringi Valley and Chumki, overa and Khiram in the Lidder Valley. The Swamp deep or barasingha (C. duvaucelli) is represented by two sub-species, the swamp-dwelling (C.d. duvaucelli) which lives in, marshy tracts of the Tarai and Duars from northern parts of the parts of the upper Gangetic plains east to Assam. The other, the hard ground living (C.d. branderi) thrives on the open grassy land of Madhya Pradesh. The population of both the sub-species together in critically low now mostly confined to the Kanha National Park in Madhya Pradesh. The Alpine must deer (Moschus sifanicus) occurs in the higher reaches of the Himalaya from Chamba Velly in the west to Sikkim in the east and the forest musk deer (M. chrysogaster) is found from Garhwal in the west to Arunachal Pradesh and Assam in the east. The family Bovidae contains 21 species, of which the under mentioned 14 are endangered. The blackbuck or Indian antelope (Antelope cervicapra) occupies open plains, while the chinkara or Indian gazella (Gazella dorcas bennetii) is basically an animal of the desert, but it also inhabits the grassy plains and scrub zones of the country. The blackbuck population is about 10,000 to 15,000 in the western and southern India, while the chinkara is over 500 in Gujarat alone. The chiru or tibetan antelope (Pantholaps hodgni) d the Tibetan gazelle (Procapra picticaudata) are both found in Ladakh while the latter extends to Kumaon and Sikkim. The gaur or Indian bison (Bogaurus) is found in the hill forests of India. The wild yak (Bos grunniens) is known within the Indian limits only from Cheng chenmo Valley in northern Ladakh. The wild buffalo (Bubalus bubalis) is now restricted to certain swampy areas of Assam and Madhyra Pradesh, its total population is around 1000, being 500 in Kaziranga, 300 to 400 in Manas (both in Assam) and about 1(0 in Bastar (Madhya Pradesh). The urial or shapu (Ovis orientalis), the Markhor (Capra falconeri) and the Himlayan ibex (C. ibex) frequent the grassy patches in the high altitude regions of Kashmir, Himachal Pradesh and Punjab, and the nayan of Great Tibetan sheep (Ovis ammon hodgsoni) in the plateau regions from Ladakh east to Sikkim on the higher reaches of the Himalaya. The serow (Capricornis sumatraensis) occurs in the thickly wooded areas of the Himalaya between c. 1850 to 3050 mt. attitude from Kashmir east to Arunachal Pradesh.

The Lagomorpha is represented by four species, of which the hispid hare or Assam rabbit (Caprologus hispidus) is a rare and elusive species found in the grassland of the Tarai and the Duars and also in the Brahmaputtra Valley.

The order Rodentia forms the largest group represented by 95 species within the Indian territorial limits, of which 11 species of flying squirrels and two of marmots are endangered, being widely exploited for their pelt as well as for their flesh. The common geint flying squirrel (Petaurista petaurista) is widely distributed throughout the wooded parts of the country but occurs in small numbers. The woolly flying squirrel (Eupetaurus cinereus) is localised in the high altitude regions of Kashmir and Sikkim. The Smaller Kashmir flying squirrel (Hylopetes fimbriatus) is confined to Kashmir. The Himalayan marmot (marmota boback) occurs in the high altitudes of the Himalayas from Kashmir east to Sikkim, while the Long-tailed marmot (M. caudata) is restricted to the high altitudes of Kashmir within the Indian limit.

As many as 25 species of Cetacea are known from the Indian territorial limits and are all endangered. The Gangetic dolphin (Platanista gagetica) is found in the river system of the Ganga and Brahmaputra. Other cetaceans are marine forms. The baleen whales (Balaenoptera spp; Megaptera sp.) and other whales and dolphin species (Kogia sp; Stenella spp; Sotalia spp; Steno sp; Tursiops sp; Lagenorhynehus sp; Orchaella sp; Pseudorca sp; Globicephala sp; Physeter sp; Ziphius sp; Neomeris sp. and Delphinus spp.) have been indiscriminately exploited for baleen for commerce, and for flesh, fat, bone and ambergris.

Birds

There are about 2100 species and subspecies of birds known in India, of which a large number of birds are greatly endangered. Below is a brief account of these species.

Of the ducks, geese and swans (family Anatidae), the pink-heued duck (Rhodonessa caryophyllacea) occurred in forest ponds and swamps along the Himalayan foothills and Terai, from Nepal east to Manipur and also appeared sporadically in places as far away as Punjab in the north and Tamil Nadu in the south. At present, it is probably extinct from India. The whitewinged wood duck (Cairina scutalata) breeds on trees in dense swampy forests of Assam; Arunachal Pradesh and Manipur. Migratory swans, mely, the whistling swan (Cygnus columbianus), the whooper swan (C. cygmus) and the mute swan (C. olor) are exceedingly rare vagrants or stragglers. They are disappearing for want of appropriate habitat.

Most of the diurnal birds of prey belonging to the families Accipitridae, Pandionidae and Falconidae live a very precarious existence due to habitat destruction, pollution hazards, poison contamination, trapping for falconry and are subject to random killing for their tendency to prey on poultry etc. Various species of hawks (Accipiter spp.) eagles (Aquilas spp.) and falcons (Falco spp.) are widely distributed but in small numbers, and they are trapped for falconry. Teh whitebellied sea eagle

(Heliaeetus leucogaster) of the Indian coastal areas and the Bay Islands are of small population size.

The Nicobar megapode (Megapodius freycinet), family Megapodidae is confined to the Nicobar group of islands within the Indian limits. Its previous abundance has been affected by habitat destruction, increased human interference and exploitation, specially for its flesh and eggs.

Phasianid birds (family Phasianidae) are in great danger chiefly from human exploitation for their flesh they have always been branded as 'game birds' Many of these are gorgeously coloured (Ornamental) and are in heavy demand from zoos and menageries fot which they are trapped.

Bamboo partridge (Bambusicola futchi) restricted to open scrub and bamboo jungles from the plains to about 2000 m altitude in north-eastern India is fast loosing its habitat due to human interference. The mountain quail (ophrysia superciliosa) known only from a few specimens taken in the late 19th century in Kumaon is now thought to be extinct. The blood pheasant (Ithaginis cruentus) occurring in the Himalayas between 2000 and 4500 m altitude from Nepal east to Arunahcal Pradesh, the western Tragopan (Tragopan melancephala) of the western Himalayas from Kashmir to Garhwal between 1000 and 3000 m altitude of the Himalaya from eastern Garhwal east to Arunachal Pradesh between 2500 and 4000 m altitude. The Indian peafowl (Pavo cristatus), although not threatened, needs to be mentioned since it is the National Bird of India.

Of the cranes (family Gruidae), the eastern common crane (Grus grus Lilfordi) is a winter migrant in the larger wetlands of northern India, the blacknecked crane (Grus nigricollis), a greatly threatened bird, breeds in Ladakh at the altitude above 3000 m and presumably also in the eastern Himalayas, the Hooded Crane (Grus monacha) is a rare winter migrant to eastern Assam, the Siberian crane or the great white crane (Grus leucogeranus) is a regular winter visitor to the protected wetlands of northern Indian plains but a decidedly diminishing species.

The masked finfoot (Heliopais parsonata, family Heliornithidae) a unique relic species, known from eastern Assam and Arunachal Pradesh in dense lowland jungle swamps and ponds. It is so rate that no authentic record of its occurrence is available for the last fifty years or so.

Of the bustards and floricans (family Otididea), the little bustard (Otis tetrax) and the houbra or Macqueen's bustard (Chlamydotis undulata) are winter migrants in northern and north-western India; the great Indian bustard (Ardeotis nigriceps), the Bengal florican (Eupodotis bengalensis) Sypheotides indica) are resident species. Their previous abundance as well as range have shrunk greatly and the birds are facing extinction.

Of the pigeons and doves (family Columbidae), the Nicobar pigeon (Caloenus nicobarica), restricted to the Nicobar group of islands, is fast loosing its breeding grounds and is confronted with slaughter by the local people.

The hornbills (family Bucerotidae) are large birds, and most species are threatened, chiefly due to deforestation resulting in a dearth of suitable nesting trees, and persecution by man for alleged medicinal values of their flesh, fat and bone, and for their casque and other derivatives used as ornaments, charm of in witchcraft by the tribal people. The most threatened of them are : the whitethroated brown horbill (Ptilolaemus tickelli) found in the hills of eastern India in evergreen and cutover scrub forests from 300 to 1500 m altitude; the rfusnecked hornbill (Aceros ripalensis) found in central and eastern Himalayas in dense deciduous and evergreen forests of Sikkim, Arunachal Pradesh, Manipur and Mizoram; the wreathed hornbill (Rhyticeros undulatus) found in the tropical and subtropical evergreen forests along the Himalayan foothills in the Duars and also in the region south of Brahmaputra from the plains to an altitude of 2500 m; the Narcondam hornbill, (Rhyticeros (undulatus) narcondomi) an endemic species in the

Narcondam Island of the Andaman group is highly vulnerable on this volcanic Island; the great pied hornbill (Buceros bicornis) of the tropical and subtropical evergreen, and most deciduous forests from the plains up to 2000 m altitude, from Kumaon east to north-eastern states, and in the Westem Ghats from Khandala (Maharashtra) kerala; the Indian pied hornbill (Anthracoceros malabaricus) of moist deciduous and evergreen forests and mango groves from the plains to 350 m altitude along the terai from Punjab east to north eastern States, and in Bihar, Orissa, Andhra Pradesh and Madhya Pradesh; and the Malabar pied hornbill (Anthracoceros coronatus) of evergreen and moist deciuous forests from the plains to 300 m altitude from Ratnagiri (Maharashtra) east to Orissa and south to Karnataka, Tamil Nadu and Kerala.

Reptiles

The Reptiles of India belong to the fo llowing groups; tortoises and turtles (Chelonia), crocodiles and gharials (Loricata), lizards (Sauria) and snakes (Serpentes). Exploitation of reptilian resources in India is chiefly for flesh and egs as food, hide, skin and shell for fancy leather articles and handicrafts, teeth and bones as charms, fat for alleged medicinal properties and venom for preparation of medicinal products. Living snake specimens are also utilized by snake charmers and there is a great demand from the zoos and menageries.

The flesh of turtle, tortoise and terrapin is widely consumed almost throughout India. They are collected from both inland and marine sources. Indiscriminate capture for trade in their flesh and shell, and collection of eggs from rookeries have resulted in many species becoming rate. Particularly endangered are the leatherback or trunk turtle (Dermochelys coriacea), the green sea turtle (Chelonia mydas), the loggerhead or olive ridely (Lapidochеyls olivacea) and the hawks bill or tortoiseshell turtle (Eretmochelys imbricata). They are all marine forms and found in the coastal areas and estuaries. All the three crocodilians found in India, *viz.* the estuarine crocodile (Crocodylus porosus) found in the estuaries of large rivers and coastal areas, including those of the Bay Islands, the marsh crocodile (Crocodylus palustris) occuring in inland swamps and larger warterbodies all over India, and the gharial (Gavialis gangeticus) found in the big rivers with islands littered with gravel, boulders and pebbles, all lead a precarious life.

All the five species of monitor lizards, namely the common Indian monitor (Varanus bengalensis) widely distributed throughout India, the desert monitor (Varanus griseus) found in the arid and semiarid zones of western, north-western and west-central India, the yellow monitor (Varanus falvescens) found in northern India in the Ganga Valley, the water monitor (Varanus salvator) found in the estuarine and freshwater swampy areas of eastern and south-eastern part of the country, and the clouded monitor (Varanus nebulosus) found in eastern India and the Bay Islads are greatly threatened.

Of the snakes, the Indian or rock python (Python molurus) found in the dense forested tracts throughout India and the reticulated python (Python reticulata) restricted to tropical and subtropical forests of eastern and north-eastern India and in the Bay Islands are largely persecuted for their ornamental skins.

Of the amphibians, the Himalayan Newt or Indian Salamander (Tylototriton verrucosus) is of much zoological interest and is found in eastern and north-eastern India. It is restricted to some small pockets in the Darjeeling District of West Bengal, Sikkim, Arunachal Pradesh and Manipur. Habitat destruction and aquatic pollution, it is feared, may greatly endanger its survival. The viviparous toad (Nectophryne tuberculosa) is a very rare amphibian from the Malabar rain forest tracts.

11

Wildlife Conservation

Introduction

Wildlife Conservation includes all human efforts to perserve wild animals from extinction. It involves the protection and wise management of wild species and their environment. Some species have become extinct due to natural causes, but the greatest danger to wildlife results from human activities. Thus, we ourselves have created this need for wildlife conservation.

The progress of man throughout has been beneficial for the human race but it is the wildlife that has suffered through the years. Invention of sophisticated weapons, industrialization, urbanization, ever increasing human population have been some of the major causes for the dwindle of our once rich resource. Hunting, clearing of forests, draining of swamps and damming of rivers for irrigation and industry this is what we appraise of man's progress. These activities have vastly reduced the natural habitats of our wildlife and many species are endangered or nearly extinct.

Man's indifference and ignorance has led him to believe that he is independent of natural laws and of natural environments and forgets that he himself is a biological being and is directly and/or indirectly dependent on the natural resources available to him. He exploits, destroys and reduces the renewable natural resources, eradicates wildlife and leads his own race to levels leading to catastrophe. All this can be put to an end by careful, long-range ecological planning.

Importance of Conservation

Alongwith the topic of conservation arises a question—Why should be care about disappearing animals ? The answer to such a question can be attributed to many reasons. Some of them are as follows :

Survival Value. Every species plays an important role in maintaining an ecological balance among the living systems of the earth. These systems must continue to function if life is to survive. Loss of any species threatens the survival of several species inclusive of man. Man's own survival depends on his willingness and ability to co-operate with the living environment. Human beings turn to nature and wildlife for spiritual, psychological, physical and cultural benefits.

Scientific Value. The study of wildlife provides valuable knowledge about life processes which has helped scientists understand the functioning and behaviour of the human body. Scientists have also gained medical knowledge and discovered important medical products by studying wildlife. By studying the effects of environmental pollution on wildlife it is possible to learn how pollution affects human life.

Economic Value. Wild species of animals provide meat for food and skin for fur. The financial value of wild species is of major importance in the economics of many nations.

Beauty. Wildlife includes a diversity of species that contributes in a special way to the beauty of nature. Such beauty enriches life and increases the enjoyment of camping and outdoor recreation.

On all continents wildlife is becoming an increasingly important recreational asset and tourist attraction. They maintain the health of the living landscape and productivity of the ecosystem.

Wildlife represents the result of many years of evolution and constitute an environmental heritage to the past, a world which once surrounded our ancestors. Moreover, why should man deny that animals as species have the right to live and coexist with ourselves ? We are responsible to the coming generation to leave as an inheritance an unspoiled world with intact living beauty and richness.

Apart from the aesthetical and ethical reasons, there are also educational arguments to preserve wildlife.

Reasons for Dwindle

Man in his zeal for progress fails to realize that his own existence is being jeopardised. "Progress", as he terms it, has made him contaminate air; pollute streams, lakes and oceans, drain swamps and wetlands, lowering the water table; rape prairies and forests; spray pesticides over water and land; kill animals excessively for immediate profit, or on false assumptions that they are noxious or for fashionable "sport". Man alone, though often in combination with his livestock, is destroying the natural environment.

"Civilization", as we apply the term, is not just cities, industrial developments, miles of highways, enormous airfields, and numerous other technological conveniences. By destroying wildlife for civilization and culture, we are only negating the meaning of the words.

If he does have to change habits, then he must be able to manage them properly in order to maintain a rich and varied wildlife. Man has to think of the dangerous trop he is setting up for himself human overpopulation.

It is man's ability to think that formulated the idea of conservation conservation is applied ecology and may also be regarded as bioeconomy. In other words, conservation is defined as sound preservation, management and sensible use of the available renewable natural resources. This decision of conservation is in our hands and man seems to be going in the right direction regarding this decision that holds the key to the future of our most valuable natural resource. Governments of several nations have passed laws and set aside national parks, sanctuaries and other wildlife reserves in an effort to save the dwindling wildlife. Such efforts have saved several species from becoming extinct. However, these efforts have to increase several fold, if not for a complete recovery, at least about 3/4 of the species threatened with extinction can be redeemed.

Classification of Scarce Wildlife

There are three major categories of wildlife that are facing extinction.

Endangered Species. They face the most severe threat of extinction and require direct human protection for survival.

Threatened Species. They are abundant in certain areas but face serious dangers nevertheless. Dangers being caused due to unfavourable environmental conditions, extensive hunting, fishing, trapping or even collecting by hobbyists.

Rare Species. They have small populations. They live in protected environments and their numbers are not decreasing.

Methods of Wildlife Conservation

This depends on the source of danger to the threatened species. In many cases, wildlife can be helped by ensuring that their environment is favourable with regards to availability of food, water and shelter. This method is called Habitat Management. It involves such action as soil conservation, good forestry practices and wildlife management.

Several species have been threatened with extinction due to destruction of habitat by human activities like deforestation for farming, pollution of air, water, land, plants and animals etc. People must set aside seperate areas like sanctuaries, wildlife reserves, parks etc. In which wildlife can survive. Hunting is another major cause for the dwindle in the wildlife and this can be prevented by enforcement of strict laws that forbid or regulate such killings.

Excessive increase in the number of one species in a given area threatens not only its own endurance but also that of the other species by consuming major portion of the available food. Such numbers must be reduced by controlled hunting or by restoring its natural enemies where they have become scarce.

Sometimes species find it extremely difficult to survive in their own once favourable environment. Such species can be protected by breeding in captivity and releasing in a protected area that satisfies the conditions favourable for survival.

A species threatened with disease can be protected by sanitation measures in a habitat.

The success of wildlife conservation depends on the knowledge of the ECOLOGY of the species which involves on understanding of the way of life of the species and its relationship with its biotic and abiotic environment.

History of Wildlife Conservation

The story of the struggle for wildlife conservation is several decades old. If all started with few foresighted men who understood the role of wild animals in living landscapes and the necessity of preserving the threatened species. Encouraged by the pioneering efforts of these men several national and international organizations were formed for conservation of the most valuable of our natural resources.

The first of such organizations was formed by a group of Ornithologists in 1922, the International Council for Bird Preservation (ICBP). The first post-War organization to be formed under the sponsorship of UNESCO was the International

Union For Conservation of Nature and Natural Resources (IUCN) in 1948. In 1961 the IUCN formed the first fund-raising organization, the International Wildfowl Research Bureau (IWRB). There are several other organizations like the Food and Agricultural Organization (FAO) of the United Nations, the Smithosian Institute and the Conservation Foundation both in Washington DC and the Fauna Preservation Society in London.

The Zoologists Gessellschaft in Frankfurt, West Germany, has raised important funds for conservation, particularly for Africa.

The IUCN though its Survival Service Commission (SSC), is dealing specifically with animals threatened with extinction. Their function is to collect data on endangered species throughout the world and to initiate action to prevent the extinction of these species. The SSC investigates the status and ecology of a species and advises governments and organizations. Such advice is often continued with direct action financed by WWF. The SSC has long maintained a list of threatened mammals, birds, reptiles, amphibians, fishes etc. called "The Red Data Book".

This conservation expansion has produced positive results in many parts of the world. Inspite of such efforts, the future remains uncertain for the world's wildlife. This uncertainity exists and intensifies with the increase in human population, destruction of wildlife habitats and spread of environmental pollution.

Wildlife Conservation in India

The Indian Wildlife possesses a rich and diverse fauna and is unique in having immense natural beauty in its different ecosystems. The fauna includes about 123 families of terrestrial vertebrates. An estimate puts the number of 400 mammals, 1200 species of birds, 350 species of reptiles and more than 29,70,000 species of insects.

The animals that make up the Indian fauna are Elephant, Rhinoceros, Musk deer (Kastura), Barking deer (Kotra), Spotted dear (Cheetal), Hog deer (Hoghiran), Mouse deer (Indian chevrotain), Swamp deer (Bara singha), Dancing deer (Sambur) Thamin, Kashmir stag. Four-homed antelope (Chau-singha), Indian Gazelle (Chinkara), Blue bull (Nilgai), Bison, Wild buffalo, Himalayan ibex or wild goat (Capra siberica), Wild boar, Wild ass, Nilgiri tahr, Lion, Tiger, Leopard, Striped hyena, Nilgiri langur (Ceropithecus John), Lion tailed macaque (Macaa silenus), Rhesus money (Maaca mulatha), Hanuman monkey (maaca mulatha). Hanuman monkey (Semnopithecus entellus), Indian giant squirrel (Ratufa indica), Porcupine, Pangolin, Peafowl (Pavo cristatus), Junglefowl, Patridge, Quail Great Indian bustard (Choriotes nigriceps), Duck pigeon, Sand grouse, Stroks and Egrets, Pelicans, Eagle, crane, Owl, Hombill, Crocodiles, Gharials, Lizards (Uromastic) and about 216 species of snakes.

Endangered Species of India

The IUCN maintains a Red List of threatened species. This list is updated annually. A 1986 list contained the following species that were threatened with extinction.

These species face the most serious threat of extinction and require direct human protection for survival.

Order	Family	Zoological Name	Common Name	Status	Habitat
1	2	3	4	5	6
	Class Mammalia				
Primates	Cercopithecidae	Macaca silenus	Lion tailed macaque	E	Southern Western Ghats
Lagomorpha	Leporidae	Caprolagus hispidus	Hispid Hare	E	Assam
Cetacea	Platanistidae	Platanista indi	Indus dolphin	E	India
Carnivora	Viverridae	Viverra Megaspila civettina	Malabar spotted civet	E	Southern Western Ghats
	Felidae	Panthera leo persica	Asiatic lion	E	Gir Forest
		Panthera tigris	Tiger	E	India
		Panthera uncia	Snow leopard	E	Himalayas
Proboscidae	Elephantidae	Elephas maximus	Asiatic elephant	E	South NE India

Perissodactyla	Equidae	Equus henionus khur	Indian wild ass	E	Gujarat
	Rhinoceridae	Rhinoceros unicornis	Great Indian Rhino	E	Assam North Bengal
Artiodactyla	Suidae	Sus sulvanius	Pygmy hog	E	NE India
	Cervidae	Cervus duvaucelli	Hungal	E	Terai Kashmir
		Cervus elaphus	Swamp deer	E	Central India
		hungal Cervus eldi eldi	Manipur deer	E	Manipur
	Bovidae	Bubalis bubalis	Wild asiatic Water buffalo	E	Assam Bastar
	Class Aves				
Galliformes	Phasianidae	Tragopan caboti Tragopan	Cabot's Tragopan	E	NE India
		melano cephalus	Western Tragopan	E	Western Himalayas
Gruiformass	Otididae	Ardeotis nigriceps	Great Indian Bustard	E	Rajasthan MP

(Contd.)

1	2	3	4	5	6
	Class Reptilia				
Testudines	Crocodylidae	Crocodylus porosus	Estuarine crocodile	E	East coast
	Garialidae	GArialis gangeticus	Gharial	E	N Indian river
	Class Insecta				
Anoplura	Haematopinidae	Haematopinus oliveri	Pygmy hog sucking louse	E	Assam

E-Endangered species.

In order to protect these species from extinction we have to know about their habits and habitats. Once we are thorough with such facts, preservation of the endangerd species can be carried out successfully. The above table shows the list of endangered species in India and a detailed description of each animal is as follows.

Macaca Silenus

Common Name – Lion Tailed Macque.

Local Name – Hind-Siah Bundar, Malayalam – Nellamanthi, Tamil -- Karungkorungoo, Arakkan.

Size —Head and Body-20-24 inches. Tail-10-15 inches (25-38 cms.) Females slightly smaller.

Distribution —Western Ghats from North Kamara, southwards to Kerala and Kanyakumari district, Tamilnadu.

Distinctive Characters —Great mane of long dark-grey or brownish grey hair from temple and cheek. Glossy black coat,

Habits —In habitats lonelier evergreen tropical forest belt between 2000-3500 ft. (610-1070 m). They are dark coloured, shy and seclusive in habit. They are seldom seen. Macaques are gregarious. They live in herds of 12-20 animals or more. The call of the male resembles that of the human voice – "COYEH" or the "COO" of a pigeon. Their movement through the treetops are slow and deliberate. Newborns are seen in September in South India.

Destruction of habitat and poaching have made this species an endangered species. Trees are the means of transport and form the habitat of this species. Hence afforestation and prevention of poaching are the methods to be adapted for saving this species.

Caprolagus Hispidus

Common Name—Hispid Hare of Assam Rabbit. Not much is known about this species except that they are the same size of the hare with a shorter, dark-brown tail. The fur is coarse and is made of dark-brown. Bristles which are paler on the undersurface of the body. The outer side of the ear is brown. The eyes are small. Hindlegs are short and stout. Their claws are long and teeth are large. They feed on bark and roots. Usually seen in Terai, Dooars and Himalayan foothills. They are very elusive.

Panthera Tigris

Common Name—Tiger

Local Names —Hindi—Bagh, Central India—Nahar, Tamil-Poolee, Malayalam-Koduwa

Size —Average length-9 ft.- 9.6 ft. (275-290 cm). Female-8 ft. 6 in (260 cms.) Average weight—400-500 lbs (180-230 kgs). Female—100 labs less (45 kgs).

Distinctive characters —Tigers have a rich-coloured, well striped, short coat. Individual variations are great.

Distribution —They are found throughout India from the Himalayas to cape Comerin except in the deserts of Punjab, Sind, and Cutch. Also found in Burma.

Habits —Tigers are found at 10,000 ft. in the Himalayas, humid evergreen forests, in dry open jungle, grassy swamps of Terai, in Sunderbans leading an amphibious life, on trees, in mud and in water. Food, shade and shelter are essential for their survival. Tigers hunt from sunset to dawn, when it is not too hot. Its prey comprises of elephant (female and young), nilgai, deer, wildpig, bears, porcupine, gaur, buffalo, panthers, other tigers and when extremely hungry feeds on fowl or fish. When game is not available, tigers attack cattle and even man. They climb trees and are excellent swimmers. Though they are heavy, they are athletic.

Mating is usually after the rains Young ones are born between February and May. Male-female relationship ends when young are born : Gestation period is about 15-16 weeks, Tigresses give birth to 2 to 3 cubs per litter The cubs begin to hunt when they are 6 months old and they stay with the tigress for 2 years. Sexual maturity is attained at the age of 3 years in the case of the tigress and 4 years in the case of the male.

Life span of a tiger is about 20 years. Hunting had greatly reduced the number in India. In 1970, a project was set in motion to prevent the degradation of the tiger population. This project was called "PROJECT TIGER". Nine sanctuaries were set aside as Tiger reserves for rehabilitation of the species. At the beginning of the project the number of tigers in India was less than 2000.

Panthera leo persica

Common Name —Asiatic Lion.

Local Names —Hindi–Sher, Gujarat–Untia bagh Malayalam–Simham, Tamil–Singham.

Distinctive Characters —Mane is lesser than of the African Lion, cost is fuller and longer tassel of hair at the end of its tail, more, pronounced tuft of hair at the elbow joints and fuller fringe of hairs on its belly. The mane is light to dark, rarely, black. Cubs are spotted or striped.

Distribution —Once found in the whole of Northwest and Central India as far south as Namada.

Now lions are restricted to the Gir forest Kathiawar, Gujarat.

Habits — Gir forest in Junagadh district covers about 500 square miles of rugged country and is composed of mainly stunted teak trees, palas, junbal, ber, patches of bamboo with an undergrowth of thorny shrubs and bushes.

Lions hunt at dusk and rest during the day. Their roars are heard throughout the night. The prey on game and cattle.

Lions do not have a particular breeding season. They are known to mate during October and November. Young ones are born during January and February. Gestation period is about 116 days. The male stays with the family and helps to defend the young and to get food for the young. Young ones are produced at intervals of at least 18 months to 2 years. A lioness gives birth to its first litter when it is 21/2-3 years old. There may be 2, 3 or even upto 5 cubs per litter.

In 1978, the number of lions in the Gir forest was less than 200.

Panthera uncia

Common Name – Snow Leopard

Local Names — Simla — Barhal he, Kunawar — Thurwagh, Kumaon — Burhel haye.

Size — Smaller than a panther with a relatively longer tail.

Head and Body-3ft — 3 in-3 ft 8 in (100-110 cm). Tail-3 ft (90 in).

Distinctive Characters — The snow leopard has a short muzzle, high forehead with a vertical chin. The colour of the coat is soft grey palling to pure white on the underside. Grey is sometimes tinged with buff. Spots are unbroken and distinct on the head, nape and lower parts of the limbs. On the body they break up into larger, paler rosettes. These rosettes are less pronounced in the luxuriant winter coat. The undersides are pale white except for a few black biotches and fur. Newly born cubs are darker than the adults.

Distribution — In India, the range extends along the whole of the Himalayan chain from Kashmir to Bhutan northwards into Tibet, Central Asia and Atlais.

Habits — Not much is known about the snow leopard. Its home is the higher altitudes of Himalayas at heights of 12,000-13,000 ft. (3660-3965 m) above sea-level, above the treeline where the rocks and cliffs form the mountaln sides. Lying by day, snow leopards hunt at night preying on wild sheep and goats, musk deer, hares, marmots and other rodents, or even larger birds. In summer, when the upland pastures are open to grazing, they follow their prey to these higher reaches and finding opportunity feed on domestic animals like goats, sheep or ponies. They establish territory near human settlements and feed on domestic animals. In winter, they move down with the animals to about 600 ft. (1830 m). The gestation period is about 3 months. Snow leopards give from 2-4 young per litter.

Uncias are persistently hunted for their valuable fur. Their soft colour and luxuriant beauty is scarcely rivalled.

Platanista Indi

Common Name --Indus Dolphin.

The indus dolphin was considered a separate species. Now considered conspecific with Platanista gangetic. They are found in the Ganges, Brahmaputra, Indus and other large tributories to the bases of hills. They are seen within tidal limits, but do not enter the sea. Its body oil is used for burning lamps and for various other purposes. Not much is known about this species.

Viverra Megaspila

Common Name—Malabar Large Spotted Civet.

This species was once very common in the coastal districts of Malabar and Travancore. It lied in the wooded plains and in the adjoining hill slopes. At present they are scarcely seen and appears to be nearly extinct. Nothing else is known about this species.

Elephas Maximus

Common Name—Asian Elephant

Local Names —Sanskrit—Hasti, Gaja, Hindi—Hathi (male, Hathni) (Female), Marathi—Hatti Tamil—Anai, Malayalam—Ana, Kannada—Ane

Size —Average height-Adult male-9 ft. (2.75 m). Female—A foot lower (30 cms).

Distinctive Characters —The Asian elephant is smaller than the African elephant. While the ears of the African elephant or enormous, that of the Asian elephant are smaller. The major distinctive feature is that of the hollow back of the African elephant to that of the rounded back in the Asian counterpart. Generally only the males have tusks, while females have a protrusion of a few inches or nothing at all.

Distribution —In India, the Asian elephant is found in the Western Ghats, southwards from Mysore, Orissa, Bihar, Himalayan range, almost all over India.

Habits — The Asian elephant is found in areas covered with forests. They are extremely adaptable and live in steamy, humid jungle or in cool elevated forests. They live in herds of 5-60 animals (usually single families). When fodder is scarce herds break up into smaller parties. Big tuskers feed away from the herd. Bulls lead a solitary life and joins a herd when he feels the urge to mate. Elephants sleep when it is hot and feed in the mornings and evenings and during nights invade open forests and raid crops. They raid and retire after midnight. They feed on various kinds of grasses and leaves. stem and leaves of wild bamboo and plantain, all species of crops, barks of some kinds of trees. A full grown elephant eats about 600-700 lbs (270-320 kgs) of fodder a day. The exact breeding season is uncertain, but is estimated at the beginning of the rains. Young appear in Autumn. The gestation period is about 20 months. Elephants give birth to only one calf at a time. Elephants have poor sight but their sense of hearing and smell are well developed.

Hunting and poaching for ivory has very greatly decreased the number of elephants in India.

Equus Hemionus Khur

Common Name —Indian Wild Ass.

Local Names —Hindi—Gharkar.

Size —Height at shoulder-3 ft. 8 in (110 cms) to 4 ft. (120 cms)

Distinctive Characters —The colour of the coat varies from reddish grey to fawn or pale chestnut. The erect dark brown mane is continued as a dark brown stripe extending along the back to the root of the tail, the lower being white.

Distribution – The Indian Wild Ass is found in the deserts of the Rann of Cutch and Peninsular India. The present population is less than 2000.

Habits – They roam in groups of 10-20, or in 2 or 3's or as individuals in the scorching sun. They are capable of remaining in the open for hours on end and run for a great distance at a speed of 30-32 miles per hour. The Indian wild ass mate in the months of August, September. The gestation period is about 11 months. Young ones appear in the months of July, August, September. They are called FOALS. They are readily tamed when young. They grow vicious and untameable animals. The present population in India is below 2000.

Rhinoceros Unicornis

Common Name —Great Indian Rhinoceros.

Local Names – Hindi-Gainda, Gargadan; Marathi-Genda.

Size – Average height of male-5 ft. 8 in (170 cms) with a girth of 11 ft. (335 cms) behind the wither.

Distribution – The Great Indian Rhinoceros was formerly extensively distributed throughout the Gangetic plain, but is now restricted to the Dooars and parts of West-Bengal and Assam.

Distinctive Characters – The skin is divided into shields by heavy folds before and behind the shoulder and in front of the thighs. The fold is not continued across the back which is a distinctive character in the Indian species. Another distinctive character is that the hom is smaller in the Indian Rhinoceros.

Habits —Rhinos prefer swamps and grasslands but are also found in wood jungles and ravines of low hills. They lead a solitary life. Rhinos feed on grass. They breed through out the year. Their Gestation period is 16 months. The young ones are born as heavy as 60 kgs and are about 105 cms in length. The female Rhino acquires sexual maturity at the age of 4 years while the male takes 7 years to attain sexual maturity.

The Indian Rhinoceros is a fast diminishing species. They are hunted for sport and their skin is used for making leather products. Their horn is supposed to have medicinal value.

In India, they are found in majority only in the Kaziranga Sanctuary in Sibsagar/Nowgong district in Assam. The Bengal Rhinoceros Preservation Act of 1932 and the Assam Rhinoceros Preservation Act of 1954 were fruitful in protecting the animal and by the end of 1978 the population of the Rhino in the Kaziranga Santuary was estimated to be around 800.

Sus Sulvanius

Common Name —Pygmy Hog.

The Pygmy hog is scarcely 10 inches high (25 cms). They inhabit the forests of the Himalayas in Sikkim, Nepal, Bhutan and Assam. Their habits are similar to that of the wild boar. They live in herds of 5-20 in number. They are nocturnal by nature and rarely seen. Hence nothing much is known about this species.

Cervus Duvanceli

Common Name —Swamp Deer.

Local Names —Hindi—Barasingha, Maha, Central India—Goinjak (male), Gaoni (female), Sal samar, Nerwari.

Size —Average height 54 in (135 cms) at the shoulder, Average weight 370-400 lbs (170-

180 kgs), Average length of the horn – 30 in. (75 cms) with a girth of 5 in. (13 cms) at mid-beam.

Distinctive Characters — The coat is almost woolly and has colour shades from brown to yellowish brown. Stags are maned and darker in colour. Young ones are spotted. The size and shape of the antlers vary.

Distribution — Swamp deers are found in Terai, Uttar Pradesh, Assam, Sunderbans, Madhya Pradesh.

Habits — They live on marsh land, seldom out of water. They are also found on grassy maidans. They are highly gregarious by nature and less nocturnal. Swamp deers feed till late morning and in the evenings, taking rest during most of the hot day. Their eyesight and hearing are moderate, but their sense of smell is well developed. They are an endangered species and strict protection is necessary to prevent their total extinction. The total number in Central India is less than 200.

Cervus Elephas Manglu

Common Name – Hangul.

Local Names – Kashmir – Hangal, Hanglu (Mate), Minyamar (Female) Hindi – Barasingha.

Size – Average height – 40-50 inches (120-125 cms) at withers. Average weight-400 lbs (180 kgs). Average length of horn-40 inches (100 cms).

Distinctive Characters — The coat is light to dark brown fading to a dinghy white on the lips, chin, underparts and buttocks. The white rump patch is divided by broad median stripe extending to the base of the tail. Fawns are spotted.

Tribution — Northern side of the Valley of Kashmir and some adjacent valleys.

Habits — The Hangul is basically a forest animal and leads a solitary life or live in small parties. They roam around in search of food. They give birth to their young ones in late May. They spend their time in summer at heights of 900-12,000 ft. (2750-3680 m). They shed their antlers before they migrate and grow new ones. This is a very regular feature among this species. In winter they descend to lower levels and feed on fresh grass. Another of the endangered species that requires direct human protection.

Cervus Eldi Eldi

Common Name — Manipur Brownanlered Deer.

Local Names — Manipur — Sanga, Burma — Tamin.

Size — Average height 4 ft. (120 cms) at withers, hinds are smaller. Average length of antler — 42 inches (107 cm). Average length of head — over 35 inches (90 cms).

Distinctive Characters — The coat is coarse and sparse. The stags have a dark brown, or black coat. Hinds are light brown or fawn. Young ones are spotted. Antlers are almost erect.

Distribution — Manipur.

Habits — The Manipur Deer avoid hills and heavy forests. They prefer open scrub jungle or flat undulating land between rivers and hill ranges. They live in small herds. Their main food comprises of grass on which they feed in the mornings and evenings. Their speed and their sharp eyesight protect them from their predators. Young ones come in the month of October. Gestation period is about 239-256 days The present number is only about 100 most of which are restricted to Keibal Lanjao Sanctuary on the shore of Logtak Lake in Manipur. This is the reason that they were given the name Manipur Deer.

Bubalis Bubalis

Common Name – Wild Asiatic Water Buffalo.

Local Names —Hindi – Arna (male), Arni (female), Marathi —Jungli mhais, Kannada—Kadu Kano Kartee.

Size —Average height of the Bull-5 ft. 6 in. (170 cms) at shoulder to 6 ft. Average weight —2000 lbs (900 kgs) or more.

Distinctive Characters —The wild buffalo is sleeker, heavier had more robust-looking than their tame counterpart. Their body is slaty black in colour. The newborn are light coloured, almost yellow in colour. Their horns are curved or spread outwards. The horns may be flat, triangular in cross-section, the horns of the cows are less massive and vary less in shape.

Tribution —They are found in the grass jungles of Terai and the plains of Brahmaputra in Assam, Orissa and Bastar District of Madhya Pradesh.

Habits —Tall grass jungles with food, shelter and water to lie in, form the natural habitat of the wild buffalo. They live in small herds. They feed chiefly on grass. Feeding is done only in the mornings and evenings and sometimes in the night. During the day they lie down in the shade. Wild Buffaloes are bold and most savage by nature. They invade cultivation in search of food. Bulls may attack without warning or provocation.

A cow with a new born calf is very dangerous. The chief enemies of this species are man and tiger. Their sense of smell is well developed and they posses good hearing. They have moderate eyesight. Calves appear in tfe months of March, April or May. Wild Buffaloes sometime mate with domestic ones.

Details about only a few species have been mentioned as information about the rest has been hard pressed to find. The above facts have been mentioned only as an example as to how one can proceed with collecting data about an endangered species for the purpose of conservation.

Modes of Wildlife Conservation in India

In India, several laws have been passed and sanctuaries and national parks have been established for the protection of the dwindling wildlife. India was probably the first country to enact a Wild Life Protection Act. The Wild Birds and Animals Protection Act was passed in 1887 and repealed in 1912. For game protection in the states, in 1927, the Foreet Act XVI was enacted. Indian Board for Wild Life was established in 1952 and this was followed by setting up of Wild Life Boards in different States in India. In 1972, a new Wild Life Protection Act was passed. Under this Act, possession, trapping, shooting of wild animals alive or dead; serving their meat in eating houses ; their transport and export are all controlled and watched by special staff (Chief Wild Life Warden and authorized officers). This act prevents hunting of femals and young ones. Under this Act the threatened species are absolutely protected and the rest afforded graded protection according to their state of population size.

In India, nearly 200 sanctuaries and national parks have been established for wildlife management. Sanctuaries are places where the killing and caputuring of any animal is prohibited except under orders of the authorities concerned.

National Parks are set up for preserving flora, fauna, landscapes and historic objects of an area. Some well known wildlife sanctuaries and national parks of India are :

Kaziranga Wild Life Sanctuary

It was established in 1926 in the district of Sibsagar, sub-division Jorhat (Assam) on the southern bank of the Brahmputra river. It covers an area of 430 sq. kms of forest grasslands and swamps and supports a fauna of 700 Rhinoceros, besides elephant, wild buffalo, bison, tiger, leopard, sloth bear, sambhar, swamp deer, hog deer, barking deer, wild boar, gibbon and birds like pelican, stork and ring-tailed fishing eagles.

Manas Wild Life Sanctuary and Tiger Reserve

It is located in the district of Kamrup in Assam, covering an area of 540 sq. kms. and is situated at an altitude of 80 metres. The River Manas runs through this sanctuary. It contains the following wild animals : tiger, panther, wild dog, wild boar, rhinoceros, gaur, wild buffalo, sambhar, swamp deer and golden langur.

Jaldapara Wild Life Sanctuary

It is situated in theJ alpaiguri district of West Bengal and covers an area of 65 sq. kms of grassland. Wild life fauna includes animals like rhinoceros gaur, elephant, tiger, leopard, deer and a variety of birds and reptiles.

Palamau National Park

It is situated in the Daltongunj district of Bihar and has an area of 345 sq kms. The fauna of this include tiger, panther, sloth bear, elephant, chital, gaur, nilgai, tiger, chinkara, chow-singha and mouse deer. The Flora is thick tropical forests.

Hazaribagh National Park

This national park was established in 1954 in Bihar. It has an area of 184 sq km of thick tropical forests. The typical fauna of this park includes wild boar, sambhar, nilgai, tiger, leopard, sloth bear, hyena and gaur.

Simlipal National Park

It is situated in the district of Mayuri Bhanj in Orissa and has an area of 2750 sq. kms. It is covered with dense sal forests and has been chosen for Project Tiger. Its typical fauna includes tiger, elephant, deer, pea fowl, talking mynas, chital, sambhar, panther, gaur, hyena, and sloth bear.

Chilka Lake

This largest inland lake has an area of 1000 sq. kms. and is about 1000 kms from Bhubaneshwar (Orissa). Its fauna includes waterfowl, duck, cranes, ospreys, golden plover, sandpiper, stone curlews, flamingoes, etc.

Kolameru Bird Sanctuary

This is a small bird sanctuary near Tadepallegudam in Andhra state. It is a breeding place for pelicans and many marine birds visit this place.

Vedanthangal Bird Sanctuary

This small-sized but very old sanctuary extends over a lake of about 0.30 sq. kms, 85 kms from Madras. Many migratory birds regularly visits thist emporary lake. The regular visitors of this lake are birds like spoon bills, open billed storks, egrets, ibis, cormorant, darter, grey heron, pelicans, snipes and dab chicks.

Guindy National Park

It is situated near Madras city and has mainly chitals and black bucks. A few albinos of black buck are also found here.

Point Calimer Wild Life Sanctuary

This is situated at the southern tip of the Thanjavur district of Tamil Nadu abutting the Palk Strait. Its back water and lagoon is visited by flamingoes and pelicans. It is close to the Vedaranyam forests which has a fauna of numerous black bucks chitals and wild boars.

Mundanthural Sanctuary

It is situated in the Tirunelveli district of Tamil Nadu and was established in 1964. It covers an area of 520 sq kms. The Tamaraparani river flows through the evergreen forests of the sanctuary. Panther, tiger, sambhar and chital form the major fauna of this sanctuary.

Periyar Wild Life Sanctuary

Situated in the state of Kerala, this sanctuary covers an area of 777 sq. kms. It was established in 1949 around the artificial lake which arose behind the dam built across the Periyar river in 1900. This sanctuary of great scenic beauty supports a fauna of wild elephants, gaurs, leopards, sloth bear, sambhars, barking deer, wild dogs, wild boars, black Nilgiri langur and water birds like grey hornbills, egrets.

Mudumalai Wild Life Sanctuary

This wildlife sanctuary was established in 1940 in North Western part of Nilgiris in Tamil Nadu. It is known for its rich forests and diversity of fauna that includes wild elephants, gaur, sambhar, chital, barking deer, mouse deer, four-horned antelope, tiger, panther, bonnet monkey, langur, giant squirrel, flying squirrel, wild dog, jackal, wild cat, sloth bear, porcupine, pangolin, flying-lizard, monitorlizard, rat sanke, python and various birds.

Ranganthitto Bird Sanctuary

It covers an area of 166 sq kms and includes a series of islands in the Cauvery river 15 kms of the Bangalore-Mysore road, near Shrirangapatnam. Its avial fauna includes open bill stork, white ibis, egret, spoon bill, wild duck, pea fowl, night heron etc.

Bandipur Wild Life Sanctuary

This sanctuary was established in 1941 by the then ruler of Mysore (Karnataka State). It is situated 80 kms south of

Mysore city enroute to Ootacamund. It has an area of 874 sq kms and is at an altitude of 1454.4 metres. Its forest is very thick and has plenty of rain fall. Its wild life fauna includes plenty of gaurs and animals like elephant, leopard, sloth, bear, wild dog, chital, panther, barking deer, porcupine and langur.

Cotigo Wild Life Sanctuary

It is located in South Goa and has an area of 105 sq. kms. It has wet evergreen forest of Five and Berla trees and supports rich fauna of the following animals, gaur, sambhar, chital, hog-deer, barking deer wild boar, hyaena, panther, leopard, jackal, otter, porcupine, and birds like parakeet lorokeet, woodpecker, kingfisher, bulbul jungle fowl, egret etc.

Bhagwan Mahadev Wild Life Sanctuary

It is located in North Goa and has an area of 240 sq kms and supports fauna almost similar to Cotigao Wild Life Sanctuary.

Sesan Gir

This famous wild life sanctuary for the Asiatic lion is situated in Gujarat state, 468 kms from Ahmedabad and 43 kms from Veraval. It has an area of 1295 km sq of semiarid country with patches of thorn scrub and deciduous trees. Its fauna includes Asiatic lion, spotted deer, blue bull (nilgai), fur-horned antelope, chinkara, striped hyena, wild boar, porcupine, langur, python, crocodiles and birds like green pigeon, partridge, rock-grouse etc.

Kanha National Park

This national park was established in 1955 in former Banjar Valley Reserve (Madhya Pradesh). This park has an area of 939.94 sq. kms and includes hilly terrain ond streams. It is 175 kms away from Jabalpur and has forests of sal trees. Its typical fauna includes animals like tiger, chital, panther, sambhar, black buck and barasingha.

Tandoba National Park

It is located in Chandrapur district (Maharashtra) and has an area of 116 sq. kms. Its fauna include tiger, sambhar, sloth bear, bison, chital, chinkara, barking deer, blue bull, four-horned deer, langur, peafowl and few crocodiles.

Sariska

This is one of the most beautiful sanctuaries situated in the state of Rajasthan near Alwar. It has an area of 800 sq kms. and has dense Dohokara and Solar forests. Its fauna include tiger, leopard, spotted deer, jungle cat, four-horned antelope, langur, porcupine, hedgehog, peafowl etc.

Bharatpur Bird Sanctuary (Koe ideo Ghana)

It is located in Bharatpur district in Rajasthan, has an area of 29 sq. kms and harbours all kinds of indigenous birds like nesting water birds, water side birds and migratory birds. More than 328 varieties of birds including cormorants, spoonbills, white ibis, Indian darters, egrets, painted storks, open billed storks, great black necked storks, etc. Many migratory birds like ducks, geese, siberian cranes, etc. regularly visit this sanctuary. Drier parts of this marshy sanctuary have animals like spotted deer, black buck, sambhar, blue bull, wild boar and python.

Sultanpur Lake Bird Sanctuary

This small sized (2 sq kms) bird sanctuary is located in Gurgaon district (Haryana) about 30 kms from Delhi. Its typical avian fauna includes crane, sarus, spot-bill, rudyshell and drake.

Shikari Devi Wild Life Sanctuary

It is located in Mandi district in Himachal Pradesh. It has an area of 213 sq kms and has the following animals : snow leopard, flying fox, black bear, barking deer, musk deer, chakor, partridge, etc.

Bir Motibagh Sanctuary

It is located near Patiala in Punjab. The fauna includes black buck, blue bull, hog deer, hare, jackal and birds like peafowl, partridge, sparrow, babller, myna, parakeet, pigeon, dove, etc.

Dachigam Wild Life Sanctuary

It was established in 1951 in Kashmir, 26 kms away from Srinagar. It has an area of 89 sq kms and has two levels : upper Dachigam at 3692.3 metres altitudes and Lower Dachigam at 1846.2 metres altitude. It mainly preserves hangul or Kashmir stag, musk deer, leopard, black buck, black bear, brown bear and babboon.

Corbett National Park

It is one of India's most famous wild life sanctuaries and was constituted in 1935 as the first national park of India. It is situated between Nainital and Gharwal districts in Uttar Pradesh. It has an area of 525 sq kms and is located within west-south bend of the river Ramganga. It supports a rich and diverse fauna of the following : tiger, panther, sloth bear, hyaena elephant, blue bull, swamp deer, barking deer, Indian antelope, porcupine, birds like bulbul, wood pecker, barbet, babbler, and bee eater, and reptiles like crocodile, python, etc.

Shivpuri Sanctuary

It is located in Madhya Pradesh and is an asylum for tigers.

Annamalai Sanctuary

This sanctuary was established in 1972 in the southern part of Coimbatore district in Tamil Nadu. It has an area of 958 sq. kms and supports rich fauna of animals like elephant, gaur, sambhar, spotted deer, barking deer, nilgai, Nilgiri langur, Nilgiri tahr, lion-tailed macaque, tiger, panther, sloth bear, porcupine and pangolin.

These are some of the major sanctuaries in India. There are several smaller sanctuaries and national parks that house some of the important species. A sanctuary or a national park may be large or small, but planning should be based on a scientific background. To provide the basic needs for the wild life the following steps may be taken as far as possible for conservational purposes.

1. The forest inside the sanctuary/national park should be left unexploited as far as possible. It would be an ideal environment for the wild life to get the fullest measure of protection, and in such preserved conditions of life they can multiply freely.

2. If total unexploitation is not possible, the cutting operations of the forest should be in small units, well-distributed over the forest. Frequent light cuts are generally better for wild life than infrequent heavy cuts.

3. The forest should not be planted with or allowed to grow only a single species of plant. Further, fruit-bearing shrubs and trees should be encouraged wherever possible and the natural openings of the forests should be preserved.

4. Poaching needs to be ruthlessly prohibited in the sanctuaries in particular and generally in other areas outside them, where wildlife still exists. Mobile armed units should be attached to all sanctuaries to patrol regularly the entire sanctuary to stop illegal activities.

5. The water holes and stream sides should be managed and guarded carefully for the benefit of wildlife. Such critical areas may be of small size, but their importance is disproportionately great because they supply important elements in wildlife ecology for larger areas during hot months of the year.

6. Annual burning of the grassland in most of our sanctuaries has become an integral part of management, since not burning these might bring about a lack of suitable fodder during the hotter months. Controlled burning has been found to increase forage and to preserve

organic material in the soil Such burning may be done in blocks with unburned areas in between, so that areas are burnt in rotation. The burning is to be confined to definite blocks by means of five-breaks. Burning at night is favoured unless it is likely to extend the fire beyond the desired limits.

7. Provision for dust bath for animals should also be made by leaving some exposed soil for their dusting. These spots may be prepared in the form of small mounds elevated a few inches from the forest floor to allow for drainage and to dry quickly in wet months.

8. Some artificial salt-licks are also to be provided in the sanctuaries since the wildlife need them at regular intervals for their normal health. Such salt-licks are necessary for animals especially in areas where natural salted clay is scanty.

9. Concentrated grazing by domestic live-stock is to be prevented as wildlife can never compete with domestic stock under ordinary circumstances and domestic animals may be responsible for transmitting diseases like foot and mouth diseases, rinderpest, surra (sleeping sickness), heamorrhagic sepicaemia, arthrax etc.

10. Cultivation near the sanctuary should be avoided as a rule as it may have an indirect effect on the wildlife. The chemical control of certain pests is often desirable for better yield in agriculture, but these pesticides can present ecological hazards for wildlife.

11. Scientific studies of the sanctuary wildlife particularly of threatened species, by qualified persons, should be encouraged in order to improve the status of the animals even by breeding them in captivity to rehabilitate them in suitable habitats.

Most wildlife could be saved by sufficiently rigid protection of their natural habitat, combined with controlled exploitation of species which are of economic importance.

Some wildlife may be driven away from their natural habitats so that their requirements are not in conflict with human interests. The only possible course for maintaining their viable number is in national parks, sanctuaries, game reserves, where they could thrive unharmed.

In the case of the most endangered species which have reached a poins of extinction in their favourable habitat, their survival may be ensured through the maintenance of breeding stock in zoological parks in suitable parts of the country.

Man, An Endangered Species

"Man is part of the vast web of life and cannot escape the natural consequences of his actions .. The sheer power of the population and technological revolutions may make man himself an endangered species in many parts of the world".

— *Stewart Udall*

The present day presents Man as the most successful species in the world. The ability to think, a sixth sense, is what places the species of Homo sapiens at the top of the ladder. It is this ability that has assisted Man to become the most dominant form of life on earth. In the game of survival, Man has come out in flying colours, this being possible only by his dependence on the environment around him. Initially, this progress of Man was very slow, but with time it has increased to a dizzying pace, this being made possible by the explosion of scientific knowledge and its application to every phase of modern life.

Man's dream for the future is machine oriented civilization. This coupled with his greed is definitely going to have an adverse effect on his natural environment, soul and spirit.

An endangered species is one whose population is decreasing In Man's case it is the ever increasing population that is making him an endangered, species. There is a general belief – more people more technology – which if followed blindly, as is being done, is to flirt with ultimate disaster.

The result of progress and technological advancement is the pollution of water, land, air, removal of forests and other natural resources beyond redemption. All these activities tilt the ecological balance on which all life depends.

Each species is a part of a food chain and man is no exemption. Thus preservation of wildlife forms a major decision in the fate of man Man's attitude towards nature only suggests that he is not ready to believe this fact. Unless this attitude changes, all one can decipher is that man is paving roads leading of extinction.

President Johnson of the United States of America in an address to the Congress in 1968 observed that - "Conservation's concern now is not only for Man's enjoyment but for Man's survival". Modern Man, if he is to survive has to improve his relationship with nature. If he succeeds - he must, he will - all wildlife too will have a better chance.

The Future

The future of our wildlife looks very bleak. If wildlife is not preserved then this world would be a paradise lost, to put it in better terms, a paradise destroyed. As Man's life becomes more and more urbanized, mechanised and machinised, the habitat for wildlife is being devoured and wild animals are losing their place on our globe. Preservation of habitat is the most important criteria in conservation of wildlife.

With the ever increasing human population, the Idea of conservation can be scarcely brought about. It is a habit for nature to eliminate anything that exceeds a limit and Man, the most powerful and highly developed of living creatures, fails to recognize the needs for his own existence in the future.

Nature and wildlife are of definite economic value and components of an environment of which Man is a part and on which he is directly or indirectly dependent. Man enjoys music, poetry and paintings, all he has to realize is that the flora and fauna are the music, poetry and paintings created by nature. Even more, they form a part of the food chain which is a guarantee for the continuity of life on earth.

Very few individuals in this world have time for recreation, leisure or intellectual interests. The rest of them think wildlife conservation is a luxury. Each of them is too busy trying his own problems for him to even consider the idea of conservation. Starving illiterates have time to think only of ways to earn their next meal and not about conservation, which for them is non-functional luxury.

Human overpopulation and illiteracy are two of the major problems that bring about the tragic destruction of environment and animal exterminations in the world. For every extinct or endangered species or subspecies of animals, two of the major factors have been the destruction of habitats and excessive cropping by man. Connected with this depletion is the lack of knowledge about the essential needs of a species for survival, its needs for habitat, and its reproductive capacity.

Several organizations are painstakingly trying to bring about a successful turn-about regarding the concept of conservation. They educate and advise on the need for conservation of our depleting natural resources. All these organizations are run by men who are devoted and work their hearts out for saving habitats from destruction and species from extinction throughout the world. They fight against social, economic, political and religious problems in different parts of the world.

It all looks like a hopeful sign that the future of the world's wildlife is of growing concern to more and more human beings. But this is a race against time. Will people and governments be aware of this before too much is lost ?

The future of world's life is still a big question ?

12

Systems Approach to Environmental Problems

Introduction

One of the great contributions of the space program has been the development of the concept of a complete systems approach to the solution of complex problems. The value of this approach has been proven. Coupled with the use of computers alternatives may be explored and critical points for decision identified.

Man's environment has suffered severe degradation because of failure to evaluate problem areas on a total systems basis and because of unforeseen consequences. Rachel Carson called attention to instances of unforeseen results from the use of chemical pesticides. Biological concentration of DDT illustrates the interactions in the biological systems. The wide use of non-degradable detergents dramatically demonstrated channels of pollution and also resulted in objectionable foaming.

The present handling of solid wastes in urban areas reveals many deficiencies in failing to consider the handling in the dwelling, collection and transportation, and the disposal, recycling, or reclamation as a complete system. The disposal method, incineration, hog feeding, or land filling, may dictate the segregation and handling at the dwelling. Economic, legal, and political factors, and public attitude may be decisive in selecting a method of disposal.

The collection of all sewage at a single plant may make possible the construction of a more efficient treatment plant, but the increased absolute waste load, even after treatment at the high efficiency, may be more than the stream can assimilate. Discharge of the wastes at several points, even at lesser efficiencies in removal, might have been possible with no difficulties.

Failure of highway planners to consider noise, air pollution, and social disruption from highway alignment has resulted in citizen action to block proposed "improvements".

How many one avoid such pitfalls in environmental planning the basic land use pattern which is adopted, or allowed to develop haphazardly, is a significant determinant for the future needs and problems of the area. The livability and character of development is related to the density and distribution of the population. One such things as to type of dwelling (individual detached homes or high rise apartments', transportation system, allocation of land uses, etc. have been allowed to develop, major changes are difficult and expensive to accomplish.

Environmental planning should be considered at the time of land use planning and development. The Department of Agriculture can supply data on soils which will reveal areas most suit, able for parks, for residential lawns, or for absorbing the liquid effluent from septic tanks. Geological surveys will show the suitability for foundations, under-ground water resource potential, and areas suited for landfill waste disposal. Now subdivisions in urban areas seldom consider these factors and do not make adequate provision for solid waste and sewage disposal. In densely populated areas, sewage collection and disposal should be provided initially. Septic tanks sometimes fail after a few years of use by city dwellers accustomed to sewers. Solid waste handling in high rise apartment and office buildings requires as much care in engineering planning as climate control, structure, and other factors.

Functional use of spaces inside and outside the dwelling unit for satisfaction of human needs, both individual and family, must be considered as an interrelated system. Spaces for individual privacy, family activities, and community social interaction are required. Stimulating variety is essential to human development and preservation of individuality.

Frequently, in the past, aesthetics and environmental problems such as noise and vibration, traffic, air pollution, and waste disposal, have been neglected by industry. As a consequence, past practice has been to separate industrial areas by buffer zones. The result has been tedium and stress of long travel time to work and increased pollution from automobiles. The imaginative approach to a pleasing architectural design of modern industrial buildings provision of landscaped open space, and the solution of the environmental problems makes it possible to consider an integration of commercial and industrial development in the residential area. Regid performance standards are essential in such circumstances.

Waste disposal facilities have been similarly shunned as neighbours. Landfills, incinerators, and sewage treatment facilities have given problems where operation was faulty. With attractive architectural treatment and proper operation, adequate facilities become more acceptable.

Analysis of an environmental problem on a systems basis will result in a better solution of that problem. However, a piecemeal attack on one problem after another will not necessarily result in a total environment which maximizes benefits to man's health, There is interaction between areas. A frequent illustration is that of an industrial hygiene problem. It may be solved by installing an exhaust system, only to create an air pollution problem. The air pollution problem is solved by installing a liquid scrubber, only to create a water pollution problem. The water pollution problem is solved by precipitating the pollutant. Now, what does one with the solid waste ?

The land, water, and air resources on earth are limited. As population increases, pollution problems are intensified. As the magnitude of pollution potential increases, higher and higher efficiencies in treatment are required to maintain the status quo.

Soon, 100% recovery of pollutants may become necessary. Under such circumstances, the prevention or elimination of waste becomes critical. Measures to reduce the amount of the waste are next in importance. The irreducible wastes remaining would present no disposal problem if they could be recovered and recycled or put to a useful purpose as raw material or to reclaim land as in the case of strip mines. The interrelatedness of the various areas of the environment is again shown by the necessity to consider the water pollution potential of land reclamation projects using waste products.

It has been shown that the evaluation and solution of environmental problems must consider biological interactions, human stress, land use patterns, and interrelatedness of environmental areas. In addition, political and ecpnomic influences and consequences may be critical and decisive factors. Many environmental problems, *e.g.* water resource development, land use, and waste (air, water, solid) management require "problem-shed", or regional, approaches to provide effective control. Political innovations may be required to effect regional environmental management.

The importance of mathematics in all branches of science, technology, engineering, socio-economics, etc. is known to everyone of you. Differential equation and integration are so common in all applied sciences that at every stage we have to turn to mathematics for solutions. The reason is that mathematics is an exact science. In contrast biology, for instance, is still regarded as an inexact science. This is because biology deals with life sciences and it started as purely descriptive science. It is inherently complex and enough discoveries have not been made to bring precision of the type mathematics has Today inexactness in biology is due to our lack of knowledge.

Environment, it is defined as "the sum of all external conditions and influences affecting the life and development of organism." This area is thus a conglomerate of all basic and applied sciences, engineering, socio-economics and law. Added to it is the fact that environment is for the living system, which naturally brings in a measure of inexactness in the area of environment.

The environment of a region, country, state, district, city, town and village is influenced by several factors, because there is a high degree of location-specificity in environment, on account of an all round diversity starting with ethnic considerations, and going over to geological, geographical, climatic, floristic, faunistic, social, economic and cultural considerations.

India is a country where at this stage, centuries are co-existing. We have both near stone-age and near space-age cultures running parallel. We still have human beings existing as part of nature in perfect harmony with it; taking from nature only what they require and nothing more and existing as a part of the eco-system, as one of the species. At the other end is the socio-economic system of the human-kind where he is a co-creator and in conflict with nature and is dominated by considerations of greed. There are all manner of integrating situations in between. We have thus problems that afflict both developing and developed countries. This is true of all phases of national development. The situation in India is relevant to both developing as also to the developed world.

Today, our thinking about environment comes from the West. Historically, our understanding of environment was holistic.

The whole universe together with its creatures belong to the Lord (Nature). One can enjoy bounties of nature by giving up all greed. Implicit in the statemeat is that no creature is superior to other and human being should not have absolute power over nature. Let no one species encroach on the rights and privileges of any other species.

This element of sustainability is ingrained in this, because the emphasis is to use nature without greed. Once an element of greed enters, exploitation starts and we cease to utilise nature for the good, the benefit and the welfare of our fellow human-beings. Sustainable development is, therefore, a concept of good and sound economic growth, a growth that can be maintained indefinitely with no or minimal damage to the environment. Good environment will ultimately beget good economics. The next question arises as to what are the factors that can promote sustainable development. These are :

(a) Population stabilisation and health care;

(b) Integrated land use planning and watershed management. Ensuring adequate amount of crop-land, woodland for grassland for food, fuel, timber and fodder;

(c) Revegetating marginal land and greening the uncultivated half;

(d) Water pollution control in rivers;

(e) Air pollution control in industrial pockets;

(f) Non-polluting renewable energy;

(g) Waste recycling and reuse;

(h) Conservation of biological diversity;

(i) Human settlements without congestion and slum improvement; and

(j) Environmental education and awareness.

The elements of the life support system, namely, air, water, land, flora and fauna and the incoming sunlight form an interacting system. Sustainability of development depends on such a complex system. Even problems like the carrying capacity of our land has number of dimensions like relationship between land area, topography, population growth, and production/productivity of food, fodder, fuel, timber, fertilizer, fibre, medicines, etc. For answering such questions, it is necessary to develop a Systems Approach. Furthermore, Mathematical Modelling is one of the important methods to resolve such complex situations. The other approaches could be PERT (Programme Evaluation Review Techniques). CPW (Critical Path Method), OR (Operation Research) and so on.

Systems Approaches have acquired a significant place in modern technological development because these help to look at the problems in totality particularly for situations where different components give an impression of not working in cohesion, thus enabling to deal with whole system involving many parameters; operational mechanisms together with available resources. All these aspects need to be taken up in an integrated approach. Systems Approach has a wide scope as it helps to resolve problems in optimal utilisation of resources at minimal cost.

For instance, water quality models can be prepared about surface water, estuaries, lakes and large rivers in selected stretches with use of water quality parameters. Equally important are models on ground water in which impact on ground water quality by solid waste disposal which leads to leaching of chemicals through rain water can be worked out with reference to rainfall pattern, soil porosity and height of ground water table. This would also include monitoring of parameters like chloride, Fe, BOD.

Similarly, air quality modelling can be undertaken with reference to pollutants from growing industrialisation. We need impact assessment of environment and therefore, short and long range transport of pollutants from industries in different terrains, meterological conditions and in different industrial complexes. Validation of models would be needed by actual monitoring for operational and regulatory purposes as also air quality standards based on effects of pollutants on human beings, animals, plants and historical monuments.

Modelling should also be attempted regarding CO_2 from fossil fuels and biomass burning together with effect of its increasing concentration on climate, urbanisation and acid rain.

A Case Study: A example where systems analysis and mathematical simulation has been useful is in European part of Russia. Here, most of the rivers flow north and account for 84 per cent of country's water resources. However, Caspian Sea Basin in south accounts for only 16 per cent of water resources but 80 per cent of the production system is located in this area.

There was a suggestion for partial diversion of northern water flow southwards to enhance productivity of the aquatic and terrestrial biological system. However, before undertaking the project, the negative consequences, of such a diversion as also estimate the maximum permissible volume of diversion had to be worked out. In this exercise, over 100 institutes were involved from 1976 to 1982. The study did not favour he project.

Application of system Analysis Techniques to environment is not new to the country and several applications already exist. These are concerned with the well known traditional areas. Nearly 50 institutions/organisations are today involved in different types of application of Systems Analysis to environmental problems. Some of the areas where systems analysis has been applied in the country are :

(*a*) *Water Supply and Water Quality Planning for* : (i) Urban water supply system (ii) Rural water supply system (ii) Water quality planning and management with reference to

A. Estuarine water quality models

B. BOD—DO coupled system models

(*b*) *Waste water Transport and Disposal System for* : (i) Drainage system in multistoreyed buildings (ii) Waste water transport (iii) Waste water treatment systems.

(*c*) *Solid waste management for* : (i) Solid waste collection system.

(*d*) *Air Quality Planning and Management for.* (i) To determine optimum stack height emission rate; (ii) To compute long term and short term variation of the distribution of air pollutants and particulate deposition arising from elevated point sources; (iii) To determine the average pollutant concentration over a specified period of time; (iv) To allocate the maximum pollution emission rates from various sources in a region.

(*e*) *Land and Water Resources development for.* (i) Degradation of soil fertility due to increase in sodium concentration; (ii) Physical loss of soil through accelerated erosion; (iii) Impact of conversion of good farm land into reservoirs.

(*f*) *Forest Eco-systems for* : (i) Tropical deforestation (ii) Erosion control (iii) Agroclimatic analysis.

(*g*) *Environmental Planning for.* (i) Planning to avoid damage due to floods (ii) Impact on environment (iii) Waste water treatment and disposal of land fills (iv) Urban and rural settlements (v) ecological modelling.

A major requirement in modelling is to have good quality data which is accurate and reliable for processing before use in models.

For this purpose we have established Environmental Information System (ENVIS) using LSI 2/10 Computer System which is hooked to the Cyber 170/720 systems of NIC. The ENVIS is geared to the needs of five types of target groups like, policy planners. decision makers, research workers, administrators and the public. Information on environment is constantly in demand. ENVIS is a decentralised system using the distributed network of data bases and ensuring integration of national efforts in environmental information collection, storage, retrieval and wide dissemination by avoiding duplication of efforts.

The Department of Environment is also the National Focal Point NFP) of INFOTERRA (An International Referral System for Sources of Information) network of United Nations Environment Programme (UNEP). INFOTERRA provides referral sources on environment form a global computerised file of about 10,000 sources in response to specific queries from both national and international users. NEP has registered more than 530 national sources in the global network. A thorough updating of these sources in the country has been completed recently.

ENVIS centres in different subjecte areas have been set up. Work in the ENVIS Centres relating to their respective subject areas of environment are referred for supplying the information,

Some specific activities taken up in these ENVIS Centres include :

1. Water and Air Pollution
2. Toxic Chemicals
3. Coastal and offshore ecology, environmental mapping in Eastern Ghats
4. Eco-toxicology; Biodegradation; Environmental Impact Assessment
5. Environmentally sound and appropriate technologies
6. Renewable energy and environment
7. Non-government Organizations, Parliamentary Questions and Media as related to environment
8. Western Ghats; Biological Diversity
9. Environmental management in Madhya Pradesh.

ENVIS will also help the proposed National Environmental Monitoring organisation (NEMO) is to be established soon. The objectives of NEMO are :

(a) To build up base line information on the state of Environment in India by monitoring the abiotic and biotic components of the eco-systems with a view to :

(i) introducing environmental dimensions in economic and technical planning of development projects;

(ii) building up capability to assess the environmental impact of proposed development projects and to propose environmental safeguards;

(iii) creating a general awareness of the need to preserve the Environment using formal and informal educational programme for different levels of the communities.

(b) Acquirement of capabilities listed above would lead to the goal of providing strong data base with sufficiently accurate information content to evolve environmentally compatible development programme with concomitant preservation of environmental quality and the productivity of natural systems. To attain these capabilities, it is proposed to set up an institutional mechanism in the form of a National Environmental Monitoring Organisation which will be entrusted with the tasks listed below :

(i) To prepare an inventory of the uses of surface waters, within the territorial limits of India, stretch by stretch and identify the highest quality use and prepare the water use maps based on the highest quality use.

(ii) To monitor the quality of the surface waters within the territorial limits of India and to develop water quality maps.

(iii) To monitor the ambient air in areas free from pollution and in areas under the influence of varied degree of pollution.

(iv) To monitor the quality of land in specific areas.

A few suggested examples are :

A. intensely cultivated

B. high pesticide used

C. high fertiliser used

D. endemic

E. material (hazardous) dumping

(c) To conduct post-mortem investigation and assess environmental impact of projects measuring changes in abiotic and biotic components of the eco-systems in the neighbourhood of the project.

(d) To monitor the abiotic and biotic components of the eco-system in specific neighbourhood of developmental projects in pre-implementation, during implementation and post-implementation stages.

(e) To promote research and development for need based environmental monitoring.

(f) To undertake nation-wide monitoring of specific pollutants like metals, pesticides, aeroallergens, special chemicals, radio activity etc.

(g) To introduce tell-tale monitoring in areas of special interests such as National Monuments, National Parks, Religious or Cultural Sites, Tourist Resorts, etc.

(h) To co-operate with other agencies in specific studies to monitor the earths resources or natural calamities and other organisations having specialist capabilities for collection and analysis of data.

(i) To co-ordinate and undertake special mass education programmes, induct mass participation and arouse a general awareness among the people for the need to preserve the environment.

The aforesaid tasks if undertaken successfully by the National Environmental Monitoring Organisation would serve the following important purposes :

(a) Provide base line reference data and requisite data and feed back for the assessment of environmental impacts for different developmental projects.

(b) Provide as a service agency the necessary inputs to experts or the Government or other users for evaluation of the current state of the environment and help the users in adopting appropriate environment management policies in some suggested areas such as :

(i) industrial siting

(ii) infrastructural development

(iii) water resources development

(iv) mineral development including fossil fuel

(v) coastal management.

(c) Provide data on assimilation capacity of sinks and thereby help user agencies in formulating measures and in assessing the effectiveness of such measures through the indication of environmental quality improvement.

(d) Serving as a data bank on environmental monitoring for all users and evolution of regional and sub-regional data-based environmental management systems.

(e) Serving as an Organisation to educate the general public against the hazards of pollution and making them participate wilfully in pollution control in spite of reverse economic implications.

(f) Providing early warning services about environmental hazards.

There would be nearly 150 Field Centres located throughout the country and would be aggregated under 32 Sub-zonal Central distributed under 5 Zonal Centres at Delhi (3), Lucknow (6), Bombay (5), Madras (8) and Calcutta (10). The Headquarter would be involved in receiving, collating, processing, interpreting and retrieval system.

Data management involving data abstraction, formulating, validation can be efficiently handled with the help of computers. Most institutions today have access to computers and many have in-house computers with facility to use outside computers also.

13

Pollution : What, Where & Why ?

Introduction

Today, the cry of "pollution" is heard from all the nooks and corners of the globe, and pollution has become a major threat to the very existence of mankind on this earth. It is the major challenge of our times. The pollution of various resources has gone to such an extent that we are unable to breath fresh air and drink fresh water. On one hand the advancements of Science and Technology have added to the human comforts by giving us automobiles, electrical appliances, supersonic jets, space crafts, better medicines, better chemicals to control harmful insects and other pests; etc., but on the other hand, they have given us a very serious problem to face : pollution.

In fact, it is not a new problem, except in dimensions which we see today. When man and other higher animals began their life on this earth, there was absolutely no sign of pollution. There was perfect balance in various natural processes. The air and water were pure and soil was fertile. The problem of pollution arose with the very civilization of man. As soon as man learned to use fire, the air pollution began. As human pollution increased, there was increase in our working sphere and with this also increased the pollution. The rapid unplaned industrial progress added to our pollution problem.

What is Pollution ?

Pollution can be defined as an undesirable change in the physical, chemical, or biological characteristics of the air, water, or land that can harmfully affect health, survival, or activities of humans or other living organisms. Note that according to this definition pollution does not have to cause physical harm. It may merely interfere with human activities. For example, a lake may be considered polluted if it cannot be used for boating activities.

The problem with defining pollution is specifying what constitutes "undesirable change" which requires value judgments. For example, chemicals spewed into the air or water from an industrial plant may be harmful to humans and other organisms living near the plant. However, if expensive pollution controls are required, the plant may be forced to shut down. Thus, workers who would lose their jobs may feel that the risks to them from contaminated air and water are not as serious as the benefits of having jobs.

Value judgments about short-term versus long-term risks can also cause controversy. Building a worldwide network of nuclear power plants may provide electricity for the present human generation, but it compells future generations to handle and store the radioactive wastes produced for thousands of years, even if nuclear power is abandoned or exhausted as an energy source.

Thus, we see that the determination of desirable versus undesirable effects of an environmentally altering activity is a very difficult and highly controversial process. The nature of tragedy, as the philosopher Hegel pointed out long ago, is the conflict not between right and wrong but between right and right. Neverthelss, although defining pollution is a difficult, controversial process, it must be done in order to control pollution.

Types of Pollution From a biological viewpoint, we can recognize two major types of pollutants : degradable and nondegradable. A degradable pollutant can be decomposed, removed, or consumed and thus reduced to acceptable levels either by natural processes or by human engineered systems (such as sewage trestment plants), as long as the systems are not overloaded. There are two classes of degradable pollutants : rapidly degradable (nonpersistent) and slowly degradable (persistent). Rapidly degradable pollutants, such as human sewage and animal and crop wastes, can normally be decomposed rather quickly if the system is not overloaded. For example, a rapidly flowing river can normally cleanse itself of some human sewage fairly quickly, but this natural cleansing process can be overwhelmed by too much raw sewage from a large city or a number of small cities or farms.

Slowly degradable pollutants, such as DDT and some radioactive materials, decompose slowly but eventually are either broken down completely or reduced to harmless levels. Some radioactive materials that give off harmful radiation, such as iodine-131, decay to harmless levels in a few minutes or hours and are rapidly degradable pollutants. Others, such as strontium-90 (produced by nuclear bomb blasts); can persist at harmful levels for decades; plutonium-239, produced by nuclear power plants, can remain at harmful levels for many thousands of years. Slowly degradable pollutants are often synthetic compounds, such as DDT and plastics. Because they are artificial substances, nature usually had not evolved processes for breaking them down. Potentially harmful pollutants that break down slowly must either be prevented from reaching the environment or controlled so that they do not build up to harmful levels.

Nondegradable pollutants are not broken down by natural processes. Examples of nondegradable pollutants are mercury, lead, and some of their compounds (Enrichment Study 5) and some plastics. Like slowly degradable pollutants, nondegradable pollutants must be either prevented from entering the air, water, and soil or kept below harmful levels by removal from the environment.

Air Pollution

Air is the most vital component, without which the question survival does not arise beyond a few minutes. A human being of normal health and height respires 22,000 times a day and the air utilized in course of such respiration is about 20 times more than the quantity of food consumed by him. Natural air at mean sea level contains nitrogen (78·09%), oxygen (20,94%) and argon (0.93%) as macro-constituents. The micro-components of air are carbon dioxide (0.0318%), krypton (0.0001%) and helium (0.00052%). Many other compounds *e.g.*, SO_2, NOx, O_3 etc. are also present in much lower concentration then those mentioned. Besides these, water vapour and dust particulates, in varying amounts and sizes, are present in air, depending upon the meteorological, topographical and industrial activities of the areas under examination.

Current of air is known as wind blowing with velocity ranging from 1 to 6 km/hr is known as light air. Air with velocity of 7 to 12 km/hr known as light-breeze and those from 19 to 44 km/hr as strong breeze. Air current exceeding the velocity of 90 km/hr is called storm and over 104 km/hr is known as hurricane. Such abnormal condition in air movement releases too much of dust particulates into the atmospheric air. Estimates have shown that, even now, hardly 12 to 15% of the total dust particulates released into the environment is man-made and the rest is by nature. Though man-made air pollution is statistically insignificant, it nevertheless, poses, great danger, because such pollution usually contains toxic elements as opposed to natural pollution. The second reason is that industrial air pollution is often around human habitation resulting in the living population being badly affected. The pollution caused by vehicular traffic is maximum in the heart of the urbanized areas, resulting in the constant exposure of the urbanites to lead, nitrogen oxide and unburnt particles of hydrocarbon. These released pollutants subsequently turn into the photo-chemically active species *e.g.*, ozone, PAN etc. which often produces smogy environment especially during winter.

Air prevails even upto 100 km above the surface of the earth and 11 km below, at the deepest point in the Pacific ocean around Phillipines. The pressure of air drops from one atmosphere at sea level to 3×10^{-7} atmosphere in the mesosphere region. The total mass of the atmosphere is about 5×10^{15} tons which is roughly one millionth part of the earth's total mass 80% of the total amount of the atmospheric air is confined in the troposphere, within 12 kms above the mean sea leve.

Meteorological Factors Responsible for Air Pollution

There are three major meteorological factors which govern the dispersal pollutants from the source into the surrounding atmosphere.

(i) Wind profiles are caused by the interactions of forces due to pressure, the earth's rotation and friction. On a windless day, the wind speed is zero at the earth's rotation and friction. On a windless day, the wind speed is zero at the earth's surface but gradually increases to a value representative of free atmosphere at a height which is of the order of 1 km, known as 'planetary boundry layer'.

(ii) The temperature is another profile which significantly affects the dispersion of pollutants. Super-adiabatic condition indicates unstable atmosphere which is highly desirable as it causes rapid dispersion of pollutants. Sub-adiabatic as well as inversion conditions are highly dangerous, leading to the accumulation of the pollutants and ultimately causing severe air pollution problems in the area.

(iii) Topography often inffluences the local wind, as for example, the mountain valley wind, where the air tends to flow down the valley at night. Similarly, the sea breeze in the coastal regions causes the stack effluents to drift over land during day and no pollution apparently appears in the night. In order to fully understand the mechanism of air pollution in the context of the roles played by meteorological factors, let us recall the different layers of the atmosphere. Of special significance are the trochosphere and troposphere regions of the atmosphere to fully appreciate the reasons of the air pollution phenomenon. Trochosphere is the region where cloud formation takes place and generally extends upto 1.5 km. The height extending upto 15 km from the surface of the earth is covered by troposphere where more than 80% of the total air of the biosphere remains confined. Within this range, the air temperature decreases with increasing altitude. This decrease in temperature with altitude is called adiabatic (no heat transfer) lapse rate and is equal to six degree centigrade for every 1000 metre height. When the temperature of air decreases faster than the adiabatic lapse rate (a condition known as super-adiabatic lapse rate), then the atmosphere is considered to be unstable. As such, the air from one altitude readily mixes with air from other altitudes which ultimately leads to the formation of looping type of plume. A fanning type of plume is caused by thermal inversion where vertical movement of pollutants is limited. The fumigation type of plume results when there is inversion above the plume and instability below. Such condition occurs when temperature instead of decreasing, increases with altitude, thus forming a very stable atmosphere. The fumigation type of plume acts as lid over the atmosphere and brings down the pollutants to ground. Thus, creating severe problem of air as well as soil and water pollution. Pollutants in air are likely to accumulate in concentration sufficiently large enough to cause such episode in the regions where several dars of low mixing height, low wind speed and no precipitation occur. The combination of these circumstances often results in air pollution episode. Studies carried out by scientists reveal that mixing height of less than 1500 metre, wind speed of less than 15 kmph (4 m/sec) and insignificant precipitation for about 4 days are very favourable conditions for creating air pollution.

Types and Sources of Air Pollution

Air pollutants are generally of two types.

(i) Primary pollutants are those which are directly produced as a result of interaction between two or more primary pollutants *e.g.* nitrogen peroxide, ozone, sulphur trioxide, aldehyde, peroxyacetyl nitrate (PAN), etc. Secondary pollutants are photochemically active and responsible for the formation for smog. Smog is a combination of smoke and fog. Smoke is an emulsion of unburnt carbon particles in air and for is the emulsion of water vapor in air. Particulates present in the atmosphere may vary in numbers from several hundred/cm^3 in very pure air to more than 10^3/cm^3. Particulate mass level ranges from 10 ugm^3 in clean air to 60-2000 ug/m^3 in the polluted air in urban areas The particles of size between 0.1-lu govern the heat balance, fog formation and provide nuclei for ice and water droplets formation. Smog has now-a-days become a general phenomenon in big cities, especially during winter, primarily caused by automobile exhaust. The automobile exhaust contains unburnt hydrocarbon which produces photochemically active compounds known as formldehyde and causes eye irritation. Lead is another air pollutant, generated from the automobile exhaust, owing to use of tetraethyllead as anti-knock material in the petrol. Mosses are excellent indicators lead pollution from aerial sources. The lead poisioning causes elevated content of amino leveulinic acid (ALA) in the urine of tne affected person. The increased concentration of bromine in the water or soil is another indirect indication of lead contamination. Evidences have shown that consumption of dust and dirt, composed of lead outfall is a major source of unusual indigestion in urban people. Dust and smoke are commonly found in air. It is the respirable dust index (RDI) *i.e.* size of dust particulates, which affect the human health. Particulates of size 3 micron (u) reach the lungs, out of the total inhaled dust, while particulates bigger tnan this size are checked somewhere in the respiratory possage of nose. Particulates of much lower size 0.1 u) are removed through expelling mechanism of the human body as cough. Out of the total retained particulates, still some get dissolved in the system. Thus, being eliminated without showing any harmful effects. Even all the permanently retained particulates are not harmful; some of them are deposited indefinitely in inert state on the lungs walls without any detectable injury. Only reactive deposited particulates cause harmful effects, but they are not equally harmful. The respirable dust producti‿n ranges from 27 to 340 gm for one on of coal mined.

In addition to the automobile exhaust, other major sources of air pollution are mining, thermal power plants and brick kilns. The latter is a pin-point source of air pollution, especially in the rural area. The intensive coal mining regions of Jharia, Dhanbad, Katras and the adjoining areas of Bihar have been identified to be the most air polluted cities nearly 15,000 tons per year of particulate matters are released in the air in Jharia coal field areas, causing complete blanketting effect when viewed aerially. Similarly, open stack coal carbonization emitts about 35,000 tons of particulate matters annually in the atmosphere. The release of sulphur dioxide through burning of coal in the boilers in thermal power stations is causing severe damage to the plants, concrete structures and also health hazard, owing to the oxidation of SO_2 into sulphuric acid. Studies conducted by Central Mining Research Station at Dhanbad (1986) indicate the mean concentration of Fe (2.48 - ug/m^3), (0.55 ug/m^3), Ni(0.04 - 0.57 ug/m^3), Cu(0.05—1 ug/m^3), Cr (max. 0.55 ug/m^3) Zn (max. 2.8 ug/m^3) in ambient air in the Jharia coal field areas.

Effects of Air Pollution on Our Cultural Assets

Neon in excess of 18 ppm, helium greater than 5.2 ppm, methane higher than 1.5 ppm. krypton more than 1.0 ppm hydrogen in excess of 0.5 ppm. nitrogen dioxide more than 0.001 ppm, carbon monoxide in excess of 0.1 ppm. and sulphur dioxide greater than 0.002 ppm in air are injurious for human health.

Ozone (O_3) in concentration of over 0.3 ppm in air causes nose and eye irritation. Its presence at a level between 1 to 3 ppm causes extreme fatigue after 2 hours of exposure and at higher than 9 ppm level, causes pulmonary edema (a disease in which excessive fluid occurs) in the lungs. Ozone concentration in 0.1 ppm for 20 days exposure causes 50% reduction in pollen germination in case of tobacco plant. Exposure of plants to ozone causes yellow spot formation on leaves which subsequently affects the rate of photosynthesis considerably. Exposure to ozone at concentrations of 0 02 ppm for 65 minutes. 0.26 ppm for 5 minutes, ppm for 3 minutes, and 20 ppm for 1 second produces cracks in rubber. PAN and aldehyde found in smog are eye irritants.

Exposure to carbon-monoxide (CO) at concentration of 100 ppm even for an hour may be fatal for human kind, probably because of its affinity with haemoglobin (red corpuscles of blood) which is about 200 times more as compared to oxygen.

Sulphur dioxide (SO_2) is one of the most potential pollutants present in the air over and around industrial areas materials like papers, textiles, leathers, stones, marbles, paintings and metasl are severely attacked by sulphur dioxide in presence of atmospheric moisture. The conversion of atmospheric sulphur dioxide into sulphuric acid is known as sulphonation. The process of sulphonation is accelerated in the presence of moisture and iron. Iron is known to be present in dust, nails etc. and moisture is present in the material itself. Cellulose which is the basic building material of paper, cotton and linen is rapidly decomposed by the process of hydrolysis. As a result, the paper and textile goods become yellowish, brittle, and ultimately loose their strength and fibrous structure. Similarly, Collagen (which is the basic constituent of leather) is also disintegrated by SO_2 and thereby, the leather becomes powdery in due course of time. Stones, particularly calcarious, like marble, which are essentially calcium carbonate, are severely damaged by sulphuric acid. Calcium carbonate is converted into calcium sulphate, resulting in huge incrustation by which the beauty of marble is changed. Paintings in which lime plaster is used, become expanded in volume and look flaky and powdery, owing to their conversion into the corresponding sulphate under the action of the atmospheric sulphur dioxide. Acidity generated by atmospheric sulphur dioxide initiates the electrolytic corrosion of metal.

The over all effect of SO_2 is summarised as below :

(i) 1 ppm for 1 hr or 0.3 ppm for 8 hrs causes plant damage,

(ii) 5 ppm for 2 hrs is a high level of concentration which causes broncho-constriction in human body.

(iii) 10 ppm for 1 hr exposure is considered a distress level for human beings.

Objects made of coper and silver are much more sensitive towards hydrogen sulphide, even in much lower concentration as compared that of sulphur dioxide. Both the metals are converted into their respective sulphides and consequently loose their artistic, ornamental and antique values.

Numerous monuments and ancient temples are exposed to the sea winds, coupled with chloride impregnated fogs at the coastal sides. The chloride ions absorbed by the stones during foggy weather are carried to the surface of the stones in dry weather, causing efforesene/ce and finally leading to the process of complete disintegration of the entity. Bronze, an alloy of copper, zinc and silver, which constitutes a large part of artifacts, is badly corroded by sea winds. The most probable mechanism of the bronze corrosion may be presented as such.

The chloride ions react with copper and form unstable cuprous chloride. The latter further reacts with environmental oxygen and moisture forming cuprous oxide, cupric chloride and hydrochloric acid. Thus, a set of chain reactions finally reflects upon the bronze objects which ultimately destroy them.

Ozone (O_3) is highly sensitive towards saturated organic materials, even at room temperature and causes their depolymerization. Resins, varnishes etc. are used in paintings and are products of saturated organic compounds. Woolen catton, paper and all other such materials that consists of cellulose are also liable to be photo-oxidised by ozone in wet condition resulting in their rapid fading of textures and colours.

The bloom in painting is caused by environmental ammonia (NH_3) in concentration sharply detactable by human nose, due to the formation of a compound known as a ammonium sulphate. Lower concentration of ammonium is though, useful, because it takes care of free atmospheric acidity and prevents other damages. Cellulose, showing, amphoteric behaviour, is hydrolysed and oxidised under both acidic and alkaline conditions. As such, the materials rich in cellulosic content are also at constant risk of being degraded in higher concentration of environmental ammonia. Furthermore, it inflames the upper respiratory passage, even in low concentration, under longer exposure.

Aldehyde (CH_3CHO) irritates nasal and respiratary tract, arsines breakdown the red cells in the blood, chlorine attacks entire respiratory tract and mucous membrane of eyes, hydrogen cyanide interfers with nerve cells, hydrogen fluoride irritates and corrodes all body passages, possibly leading to cancer.

Air Pollution Control

The safety of three M's :– Maa, material and machinery must be the ultimate object behind the measures to control the environmental pollution problems. Pollution control measures should be considered as social responsibility and obligation not only by the concerned departments, but also by all who have faith in the need for preserving the valuable natural environmental resources. The implementation of the control measures should be based on the principle of recovery or recycling of the pollutants and must be taken as an integral part of production *i.e.*, never as a liability but always as an asset. Air pollution may be controlled by physical as well as by chemical methods. Some of the most effective methos are show below ;

(*i*) *Dilution Method.* Physical methods include the suitable height and design of the stack. The minimum-height of the stack has been worked out to be 38 metres in order to ensure proper mixing of the discharged pollutants into the adjoining atmosphere.

(*ii*) *Air Cleaning.* Air cleaning device removes or recovers suspended solids, liquid and also soluble gases and vapours from air before they are discharged into the atmosphere. The commonly used air cleaning devices are : (1) Gravity settlers, (2) Cyclone separator, (3) Wet collectors, and (4) Electrostatic precipitators.

(*iii*) *Substitution Method.* The method is based on the principle to substitute the offending materials and processed by less offending ones.

(*iv*) *Plantation.* Intensive plantation in the region considerably helps in reducing dust, smoke and other pollutants. Studies conducted by the authors indicate that the trees with simple leaf, *i.e.*, pipal pakur, banyan, teak, sal, debdaru, mango, kachnar, kadamba etc. are more effective in removing dust particles suspended in the air than those of the trees *e.g.* ashoka, palas, nim, sandal and gulmohar with compound leaf. The study further shows that the dust removal from the air is best achieved by banyan (5.35 + 0.12 g/m^2 of leaf), followed by debdaru (4.56 g/m^2), sal (4.50 g/m^2), pipal (4.10 g/m^2) pakur (4.09 g/m^2) and the minimum by kadomba (3.57 g/m^2) among the trees with simple leaf. Similarly, the highest proportion of dust particles settlement from the air is ensured by ashoka (3.78 g/m^2) followed by palas (3.06 g/m^2), nim (2.92 g/m^2), sandal (2.24 g/m^2) and the least is achieved by gulmohar (1.44 g/m^2) among the trees with compound leaf. Therefore, along the road side and at places where maximum dust generation is possible, trees of simple leaf should preferably by planted. Forensuring the involvement participation, of the public plantation should be encouraged on the pattern of the social forestry. The principle of atleast one tree for each school going student, two by each teacher and three during rest of the period of life by each individual should be practised in true sense of the term. Besides, the forest conservation act of 1980 should by lup-dated and be made stringent for those who are cruel to trees and plants.

(*v*) *Chemical Methods.* This method is usually for removing sulphur dioxide from air, resulting from the burning of coal, with rich sulphur content. Several chemical methods have by now been developed to have an effective control over air pollution problems. Briefly, some of these methods are discussed below :

(*a*) *Beneficiation Process.* By this method coal is treated with lime or dolomite and processed in such a way that both the sulphur and ash content of the original coal are significantly reduced with improved yield of energy on burning. Thus, the release of sulphur dioxide is automatically reduced considerably. The other method is to fix the sulphur dioxide in she combustion chamber itself by the spraying of ammonical liquor.

(*b*) *Reduction Process.* The SO_2 emission is adsorbed by alkaline alumina, resulting in quick reduction of the former into the elemental sulphur.

(*c*) *Catalytic Oxidation Process.* Vanadium oxide is generally used as catalyst where by SO_2 is converted into H_2SO_4 or ammonium sulphate.

(*d*) *Control Measures for Smog.* Four techniques are generally used to control hydrocarbon emission from stationary sources. They are incineration, adsorption, absorption and condensation of the exhaust gases. In the adsorption method, exhaust gases are passed through a bed of a granulated adsorber where hydrocardon vapours arc adsorbed into the adsorber. The absorption method is similar to adsorption, except that the exhaust is brought into intimate contact with a liquid in which the hydrocarbons dissolve or become suspended. In the condensation method, exhaust gases are passed over low temperature surfaces where they are condensed. Liquid hydrocarbons are left behind and collected.

Control of hydrocarbon emissions from automobiles is more complicated. Several engine design characteristics favour lower exhaust of hydrocarbon emission. For example, reduction of surface/volume ratio of the combustion chamber decreases hydrocarbon emission markedly.

(*e*) Ill maintained auto-vehicles with out-dated silencers should not be allowed to play on the road.

Water Pollution

Water is relatively more prone to pollution than air because it is more viscous, having lesser moment of inertia and possessing closer molecular distance than air. Air, being available in the environment in much larger proportion as compared to water, the pollutants in air are quickly diluted and dispersed away along the down wind direction. Pollution load is thus, drastically reduced in air. The unequal distribution of water on the surface of earth and the fast declining availability of fresh usuable water per capita, per year, due to increasing population, are the major causes for concern in terms of the quantity and quality of water. Water pollution problem is more delicate and severe than that of air. The fundamental difference between the pollution of air and water is as such.

Air pollution is of widespread nature whereas water pollution remains more or less confined. Therefore, the latter is localised in nature. Air pollutants are known to travel thousand of kilometres without loosing their identity and, therefore, are not matters of immediate concern. The pollutants in water undergo many reactions and subsequently sometimes turn into hazardous forms. The study of air pollution is costly but the principles involved are relatively simpie than in water pollution studies and management.

Water may be present in rivers, ponds, lakes and in ground reservoirs. The surface water bodies get easily polluted whereas ground water reservoirs become polluted only after prolonged exposure. Among the surface water bodies, the rivers are less prone to pollution than ponds and lakes, because of the current available in the former. But once the rivers become polluted, complex and costly techniques for control of such pollution are needed. The same is true in case of ground water pollution, which is difficult to treat when and where such problems exist. So far as the effects of pollution on the living biota are concerned, they are again of conflicting nature. The rivers disperse away the pollutants immediately in the downstream and thus, become free from pollution, but the effects of pollution become established and detectable more easily in river than in ponds and lakes because the biota in river and generally less tolerant to pollution whereas the biota living in the stagnant water bodies are more tolerant to pollution.

Sources of Water Pollution

There are generally two sources of water pollution :

(*i*) *Point sources* of pollution are of definite identity with almost constant volume and fixed composition of the effluents discharged. Industrial waste waters are examples of the effluents generated from point sources of pollution.

(*ii*) *Diffused sources* of pollution are points, situated far beyond the sight having wastes of unknown composition and volume, The street run-off, agricultural run-off and other such wastes are examples of the effluents produced from the diffused source of pollution.

It is much easier to design and operate the treatment plants for correcting the effluents produced from the point sources of pollution than for the effluents produced from the diffused sources of pollution. It is practically impossible to treat the effluent of the latter because of very huge investment of money, time, labour and the lack of availability of sophisticated technologies in our country.

The potential source of water pollution in all developing or underdeveloped countries are sewage and industrial wastes. For example, out of the total wastes discharged into river Ganga, volume of sewage discharged constitutes 72 per cent whereas 25 per cent discharge is contributed by industrial wastes. In term of BOD, the pollution load contributed to river Ganga by both the pollutants is in 1 : 1 ratio. The effluents from agricultural and street run-offs join the river in much larger proportion, especially during the monsoon. The excreta contributed by nomadic people living along the banks of rivers or in the respective catchment areas are the only potential diffused source for causing the bacteriological pollution of surface water bodies in India.

(i) Sewage

Sewage constitutes of 99.9 per cent solid and 0.1 per cent liquid. 70 per cent of these are organic constituents and the remaining 30 per cent are inorganic. 65 per cent of organic portion are proteins while carbo-hydrates and fats are represented by 25 per cent and 10 per cent respectively. Estimates made by Arcievalla suggest that 6 to 12g of nitrogen and 1 to 3g of phosphorous are generated by every person every day in the form of sewage. Sewage excreted by a human being contains more than 400 different species of bacteria and viruses. About 100 viral species in sewage are known to cause diseases in human beings. Out of this, 90 per cent are contained in human sewage, 100 ml of sewage contains 10^9-10^{10} bacterial cells out of which 10^6-10^9 may belong to coliform groups and 10^5-10^6 may be of faecal origin. The Salmonella typhosa variety ranges between 10^1-10^4. Sewage has been established to contain potential fish pathogens such as Acromonas, Pseudomonas etc. Myxobacteria are also present in water, polluted particularly by sewage. Well treated sewage may still contain pathogenic bacteria and viruses, unless properly chlorinated before its discharge into any water course. Sewage, on drying after treatment, is known as sludge. Sewage further diluted by kitchen and other domestic wastes are known as sullage. Sewage may be strong, average and weak. Assessment of its strength is determined on the basis of chloride content, the ratio of BOD and permanganate value and on the basis of a combination of chemical parameters, as shown below.

1. Chloride

Weak sewage contains an average of 60 ppm chloride, sewage of medium strength shows 1(0 of chloride and strong sewage indicates 150 ppm of chloride.

2. BOD/PV (3 minutes) ratio

Weak sewage indicates BOD/PV (permanganate value) ratio equivalent to 60, average sewage near about 100 and strong sewage shows a ratio of over 170.

(ii) Potential Pollutants in Industrial Wastes

Industrial effluents generated from sugar, distillery, tannery and paper industries are accompanied by very high organic load with BOD values ranging from 2500 to 1,22,00 ppm. The paper and pulp industrial waste contains Lignin which is made up of cellulosic material and remains unattended by microbes for long, causing depletion of fishes up to as long as 40 km distance in the downstream. The wastes from oil refinery and steel industries contain the most objectionable pollutant known as phenol, causing high odour and toxicological problem in the receiving water body. Besides, the latter also contains cyanide in appreciably high concentration to produce detrimental effects on the biota in the receiving water bodies. Fertilizer industries discharge effluents with high content of ammonia (100 ppm), urea (180 ppm), phosphate (40 ppm) and sulphate (1200 ppm), which subsequently find their ways into the adjoining water courses, causing bloom of algae. Ammonia in concentration greater than 0.1 ppm in river, stream or lake is toxic to fishes. The fertilizer waste is usually accompanied by arsenic and chromium and unfortunately both of them are equally objectionable for aquatic fauna and flora. The wastes generated from alkali manufacturing industries contain mercury, which builds up 1000 times higher concentration in fishes than found in the receiving water course. The minimata disease of Japan is a well known example of mercury hazard where thousands were poisoned and hundreds died by consuming mercurised fishes. Lead is another accumulating poison which is found in wastes generated from battery, printing, petroleum and pastes processing industries. So many of trace and toxic elements, *e.g.*, zing, boron, coppor etc, are found in large concentration in the effluents produced by non-ferrous industries. The effluents generated from mining operations are either highly alkaline or acidic and as such both are injurious for the aquatic organisms. The phosphate Pyrite mining produces wastes of very high acidic nature, accompanied by as low as pH value as 1. The mining mica and coal sometimes produces acidic wastes but these are not as strong as pyrite effluent.

All the industrial wastes generated during processing and manufacturing stages must be suitably treated before their discharge either on the land or into the water course. The cost involved in their treatment is negligible in comparison to the gain achieved.

Tolerance Limit for Pollutants Present in Water

Water is used for a variety of purposes. It may be used for drinking, bathing, cooking, irrigation and industrial purposes. Its industrial use may again be of various types. It is used in boiler, laundary, power generation, breweries, dairies and in many edible productions and processings. The standard of the quality of water vary considerably in each of the above cited industries. Detail for the quality of water standards, applicable in each of these cases is not possible to present here. However, some of the important parameters for the maximum permissible tolerance limits are below :

(*i*) *Physico-chemical Standard for drinking water.* Lead, 0.1 ppm, Cadmium and Seleninm, 0.01 ppm for each, Chromium, Arsenic and Cyanide, 0.05 ppm for each, Phenol and Mercury, 0.001 ppm for each Turbidity 10 JTU, pH 6.5 to 9.2, Total solid, 500-1500 ppm. Hardness ($CaCO_3$) : 600 ppm, Magnesium : 150 ppm, Iron : 0.3-1.0 ppm, Manganese-0.5 ppm, Copper-1.5 ppm, Chloride-1000 ppm, Sulphate-400 ppm, Flouride-1.5 ppm, Nitrate as (NO_3)-45 ppm, Nitrate as (N)-12 ppm, Zinc-15 ppm.

(*ii*) *Bacteriological Standard (WHO, 1971) for treated drinking water samples.*

1. E. Coliform should be absent.
2. 95% of the samples examined should not have any coliform count throughout the year.
3. If coliform counts were found, then in no case, it should exceed 10/100 ml.

OR

Coliform could should not be found in any two consecutive samples.

(*iii*) *Maximum permissible chemical limits for inland surface water, intended to be used as raw water for public water supply and bathing ghats.*

BOD-3 ppm, pH-6 to 9, Chloride-600 ppm, Sulphate-100 ppm, Cyanide-0.01 ppm, Arsenic-0.2 ppm, MPN count-5000/100 ml.

(*iv*) *Bacteriological standard for raw water to be used as source of Public water supply.*

Relaxation has, however been granted in case the total coliform count exceeds the maximum permissible range of 5,000 100 ml for class C water sample, as under.

Not more than 20% of the total number of water samples examined should exceed the maximum permissible range for total coliform (*i.e.*, 5000/100 ml) or almost 5 per cent of the total semples may hardly exceed 4 times the maximum prescribed limit. One more condition has been added by IS. 229-1982, which states that feacal coliform count should not exceed 40 per cent of the total coliform count.

(v) *Limits for industrial effluents to be discharged into water course (IS : 2490-1974)*

BOD-30 ppm, COD-250 ppm, Total suspended solid (TSS)-100 ppm, Sodium-0 2 per cent, pH-5.5 to 9.0, Temperature-40°C. Chloride-600 ppm. Flouride-2 ppm, Zine-5 ppm, Copper-3 ppm, Nickel-2 ppm. Cadmium-2 ppm, Chromium and Lead 0.1 ppm each, Arsenic and Cyanide-0.2 ppm each, Mercury-0.01 ppm, Selenium-0.05 ppm

(vi) *Limit for industrial waste to be discharged into public sewers (IS : 3306-1974)*

BOD-500 ppm, pH-5.5 to 9 TSS-600 ppm, Temperature-45°C, Chloride 600 ppm. Sodium-60 per cent, Cyanide-2 ppm, Copper-3 ppm, Lead-1 ppm.

(vii) *Limit for the discharge of sewage into the water course (IS : 4754-1973)*

BOD-20 ppm, TSS 30 ppm.

(viii) *Maximum tolerance limit for water intended to be used for irrigation*

The suitability of water to be used for irrigation shoul d strictly satisfy the following criteria :

(*a*) Sodium Absorption Ratio (SAR)

$$SAR = \frac{Na}{(Ca+Mg/2)^{1/2}}$$

Note : The values of SAR should not exceed 26.

(*b*) Soluble sodium percentage (SSP)

$$\frac{Na \times 100}{Ca+Mg+Na}$$

Note : SSP must not exceed 60, otherwise the water would become unfit.

(*c*) Potential soil salinity (PSS) :

Cl (epm)+1/2 SO_4 (epm) where, epm is equivalent part per million which is equal to ppm/ev and ev is equivalent weight.

The water having 6 epm or more chloride is injurious for acidic plants.

(*d*) Residual sodium carbonate (RSC) : $(CO_3+HCO_3)-(Ca+Mg)$ in epm Water with RSC greater than 2.5 is unfit, and those with RSC between 1.25 and 2.5 is marginal and water with less than 1.25 epm of RSC is good.

(*ix*) *Chemical quality criterials of grounwater.* The water exceeding the maximum permissible concentration of any single or a number of constitutents, not only unfit for drinking, rather creates many problems. For example, waters having more than 400 ppm of bicarbonate, 100 ppm sulphate or 40 ppm silicate may contribute incrustation to the carrying pipes. Waters showing 2 ppm or more of iron or 1 ppm or more of manganses tend to accentuate the process of incrustration, especially when sulphate, silicate and bicarbonate are at optimum concentration in it. Similarly, waters with low total dissolved solid (TDS) and low PH value contribute to chemical corrosion whereas, the waters with high TDS, usually greater than 1,000 ppm of high pH value lead to the electro-chemical corrosion.

(*x*) ***Biological quality criteria for limnological study***

1. Biotic Index (BI)

A stream, pond of lake with a biotic index of more than 15 is clean and when it is lower than 10, is considered to be polluted.

Biotic Index=2 (number of class 1 organisms)+(number of class II organisms).

Class 1 organisms include caddis fly, may fly etc. and are pollution intolerent. The organisms belonging to class II category *viz*. dragon fly, shrimps, plenoria etc. are indicators of mid pollution Class III organisms *e.g.* mosquito larva, chironomoids, tubifex etc. are indicators of heavy pollution and their presence marks zero biotic index.

2. Sequential Comparison Index (SCI)

The determination of the values for SCI does not require much of a background of biological knowledge, rather it is done through simple technique, even by persons of ordinary skill. The mathematical expression for SCI values are as such.

$$SCI = \frac{\text{Number of runs}}{\text{Number of organisms}} \times \text{No. of equivalent species.}$$

The SCI values decrease with increasing intensity of pollution because the latter leads to a significant reduction in the number of community species at the cost of the increase in the number of individuals of a particular tolerant species. Generally, SCI value less than 5 indicates severe pollution, between 5 to 10 mild pollution whereas a value above 15 is indicative of clean water zone. For confidence limit of 90 per cent with mean difference of 100 per cent, requires a minimum of 7 to 8 times determination for each of all the samples drawn from the bottom.

(*xi*) *Algal Genus Index* (*AGI*). Score of 20 or more secured by the algae belonging to different general in water sample is indicative of organic pollution. Score lying between 15 and 19 is taken a probable indication of high organic pollution. AGI value lower than 5 is indicative of clean water zone. The presence of algae means the presence of 50 or more individuals of a single genus in one ml of a water sample.

Soil Pollution

Soil sustains innumerable microbes and a large number of plants and animals along with plenty of mineral reserves for purposeful exploitation. Soil has different meaning to different people. Chemists consider the soil like an object endowed significantly with high buffering capacity while to a geologist, the soil is a means for ascertaining the age of the earth from a large number of isotopes present in it. The soil to a limnologist is a dependable source of nutrition for the biota, and to a farmer, the soil in the form of land is a source of living whereas to a sanitary engineer, it is nothing more than dumping ground for all kinds of wastes. Above all, the soil is a living record from where originated all major ecological events of the past.

Land is a valuable but fast depleting resource with increasing population. It has been estimated that the total surface area of the earth is 3, 15, 14,640 km². About 30 per cent or 91.56,800 km² is land surface. Of this nearly 80 lakh km² forms the Antarctica, which is too cold to be inhabited by human population. Another 1 crore 44 lakh km² area is covered by desert and 1 crore 75 lakh km² is mountaneous regions (*e.g.* 80 lakh km²) are not habitable. Thus, human population is left with hardly 75.2 m km² (7 crore 52 lakh km²) of land surface to live in. Presently, there are more than 75 people living per km² of land. The per capita availability of habitable land is likely to be further reduced to less than 50 per cent of its present proportion by 2000 A.D. (Many of the highly urbanized centres of the world are faced with acute space problem. For example, there is hardly 20 sq. ft. of living space per capita available in a crowded city like Calcutta, thereby, causing a serious ecological concern. The decreasing availability of living space is not the only problem faced by the world; rather the fast degradation of land is a more serious aspect of pollution.

India is loosing nearly 6,000 million tons of top soil per year in Ganga basin alone, caused mainly by ruthless deforestation, thus leading to severe erosian problem. Statistics prepared by the Ministry of Agriculture, Government of India (Referred from the World Environment, 1985), indicate that out of 304 million hectare (mh) of land, nearly 175 mh is becoming less and less productive year after year. Such increasing trend of desertification is really devastating and needs our urgent attention towards waste land management. Besides, increased intake of pesticides fertilizer along with radioactive out-fall, is putting additional poisonous ingredients into the land surface and causing serious pollution problem.

Sources of Soil Pollution

There are numerous sources for soil pollution, but the important, among them, are the following three :

(i) Indiscriminate discharge of industrial effluents on land and in water course.

(ii) Open defecation by animals and human beings especially in rural areas of the country.

(iii) Unscientific disposal of solid waters on land.

In developing countries like India, the major source of soil pollution is of the third type, and is prevalent in both urban as well as in rural regions. The garbage generated by the community in India contains 30-40 per cent of compostable matter from which good qualities of manures can be obtained through anaerobic digestion. Indian refuse, usually contains less than 1 per cent of plastic and glass materials. The paper content of the refuse varies between 3-7 per cent. Further, among the physical characteristics shown by them, the density of the refuse varies from 330-560 kg/m^3 and the per capita refuse production in Indian cities varies from 0.15 to 0.35 kg/day, depending upon the social, economic and educational status of the people living in the community. Generally, the people with modern scientific vision produce less refuse than those with poor and dirty living habits. Based on the comprehensive survey conducted by it on the overall physico-chemical composition of the refuse, NEERI has recommended scientific conversion of refuse into compost. Accordingly, a compositing plant with a capacity of 200 tons per day of manure production has been installed at Calcutta. All other big and medium size Indian cities must follow similar practice to keep their cities clean.

Radioactive substances produced during fission and fusion enter into the life supporting environmental cycles. Nuclear fission is a process in which heavy nuclides are broken into lighter ones whereby energy of the order 10^9K. Cal is released. On the other hand, a process in which lighter nuclides combine together to form a heavy nuclide is known as nuclear fusion. The high temperature of sun is maintained through nulear fusion whereby, hydrogen atom is converted into helium. All elements released into environment do not enter into the food chain, rather many of them go into marine ecosystem and many of them are so diluted by nutrient rich terrestrial ecosystem that they hardly remain harmful. Examples of such harmless radioactive elements are Manganess-54, Iron 59, Cobait-60 and Zinc-65, those that are present in particulate or collodial from *e.g.* Parasedodymium-144, Zirconium-9 and Rhodinm 106, which get transferred into the marine ecosystem in the highest amount. In contrast, the soluble fission products such as Strontium-90, Cesium-137 are found in highest possible amount in land and plant. Further, Cesium-137 gets accumulated only in certain plants like mushrooms and lichens and hence does not pose as great a hazard as Strontium-90. The greatest hazard of Sr-90 lies in its accumulation in the human system through various plants and foods like milk, grass, vegetables etc., and creates a negative balance of calcium present in the human bone.

Food Pollution

Food ranks third in position, after air and water, for the maintenance of the process of life on this earth. Strength to fight against diseases depends on the intake and purity of the vital constituents of food, such as proteins, fats, carbohydrates, vitamins and minerals. Removal or denudation, in part or in whole of any of the nutrients from food articles, or addition of adulterants adversely affect our health. The strength of a nation is by and large dependent on the health of its population. The proverb "Health is Wealth" and "sound mind in sound body" are well known to us. Therefore, maintenance of the purity of food and preventing adulteration dy harmful ingredients are direct contributions to the national health development programme. Food adulteration act has long been enacted

(1954) by the Government of India, under which food adulteration is an offensive criminal act and is liable for deterrent punishment.

Adulteration, Pollution and Contamination of Food

International mixing of foreign substances with food materials, for undue economic gain is known as adulteration. The entry into the food chain, of undesirable substances through environmental agencies *i.e.*, air, water and soil is termed as food pollution. The production of toxins in food through microbial agencies is known as food poisoning. The degradation in quality of food materials under the microbial influence to such an extent so as to produce diseases on consumption is termed as contamination of food.

The pollutants may be of chemical and biological types. Both types of pollutants enter into the food materials either through environmental routes or by adulteration. Some of the pollutants have been reported to be naturally present in harmful concentration in some of the food materials. The potential chemical ingredients present in food, as pollutants are flourides, nitrate, pesticides, radioactive isotopes and many trace element, *e.g.*, selenium, lead, mercury, chlomium etc. Naturally present organic pollutants in some of the food are-solanine, gossypol, oxalic acid and erucic acid. Commonly used food adulterants are argemone oil, tricresyl phosphate and white oil for mustard oil, vanaspat (hydrogenated oils) for butter, and chromium or lead for turmeric (haldi). Toxins are produced in food by clostridium batulinum, cl. perfringens staphylococcus aureus, salmonella and bucillus. The detail for the above mentioned, pollutants, adulterants, food addatives and bacteriological toxins are briefly presented below.

CHEMICAL POLLUTANTS

(i) Flouride

Tea, spice, betel, gram, french beans and fishes are rich sources of fluoride. Fruits and vegetables are generally low in fluoride content. Among cereals, fluoride content is minimum in rich preparations while it is maximum in chapati made of wheat. Which contains 0.8-1.8 ppm of fluoride. A high intake of calcium, magnesiun and vitamin C is known to play an important role in preventing chronic fluoride intoxication. People, living in the regions where water contains more than 1.5 ppm of fluoride, are advised to take citrus fruits which are rich in vitamin C. Besides, they are advised to avoid spicy food because it increases the flouride.

(ii) Lead

Presence of lead in human blood in more than 40 microgram per 100 grams blood can inhibit enzymes such as ALA which are essential for biosynthesis of haem (a basic unit of blood). Soft water, paints etc., also contribute lead to our food materials.

(iii) Cadmium

It is not required by the human body even in the trace amount and therefore is completely absent in an infant. The cadmium intoxication can considerably be prevented by increasing the intake of vitamin C Cadmium burden in excess of 20 to 30 mg in the human body may create a negative calcium balance, leading to osteomalcia (softening of the bones. Multiple factures resulting from osteomalcia are responsible for the painful discase known as Itai-itai.

(iv) Tin

Oils, fats, ghee, beverages, jams, fruits, vegetables, biscuits confectionery, coffee, tea etc. are marketed and stored in consumer's houses in containers generally made of tin. Acidic food stuff are more prone to be contaminated by tin. Canned moist food *i.e.*, butter, grape fruit juice etc. on opening starts dissolving tin. For example, canned tins apple juice when first opened may contain 20.3 mg of tin per kg of juice while after 2 days, the quantity may go upto 87 mg/kg and after 4 days, upto 364 mg/kg. The food stuff and beverages normally contain tin in amounts which are usually below 100 mg/kg and are generally well tolerated. The toxic dose in food which are causes acute toxication is believed to be around 2 to 3 grams/ day for weeks together.

ADULTERANTS

(i) Argemone Oil

The argemone sects are black in colour and similar in size as mustard seeds any latter is therefore very been adulterated with the former, it has been reported that regular consumption for 1-3 months of and edible oil, contaminated with as low a concentration as 0.1% of argemone oil can give rise to epidemic dropsy In human beings. Dropsy is a disease, causing swelling of feet, thighs and face. If the consumption of adulterated oils is not discontinued, it may cause glaucoma (disease of the eye) leading to blindness or swelling of liver which may be fata.

(ii) Tricrecyl Phosphate (TCP)

TCP is practically colourless, odourless, non-volatile and stable liquid, insoluble in water but is freely miscible in vegetables oils. It is used as adulterant with the mustard oil. Consumption of oils containing TCP for at most 20 days may cause flabbly paralysis. Like parathion (a phosphorous pesticide), it may also impair the nuero-muscular control of the body.

(iii) Vanaspati (Vegetable Oil)

The fundamental difference between ghee (butter, oil, vanaspati is that the former contains saturated fatty acids whereas the latter contains such in free state. Vanaspati unlike ghee, contains trans-acid which may cause adverse effects on the consumer's health on prolonged use.

Vanaspati is not only cheaper than ghee, but also closely resembles the latter in appearance. Vanaspati is, therefore, commonly used for adultering ghee. Such adulteration can very easily be detected by the following method even without going through any complicated chemical tests as such ;

Ghee is taken in a test tube and is method in a burner or spirit lamp. One spoon full of sugar and 10 cc of concentrated hydrochloric acid are added into it. The mixture is then warmed for two minutes. Development of red colour confirms the adulteration of the ghee with vanaspati.

MICROBIAL TOXINS

(i) Botulinus Poisoning

Such poisoning is caused by colostridium botulinum (an anaerobic spore forming organism) in improperly sterilized preserved foods which have a near neutral or slightly alkaline condition. The poison caused by cl. botulinum is most potent because ingestion of even minute quantities would result in a severe syndrome that may be fatal within 2 to 10 days.

(ii) Staphylo-coccus Poisioning

Stapylo-cocci bacteria grow in cream, puddings, sauce and other starchy preparations. Besides these, milk, egg and meat are also prone to their attack, especially during summer, warm foods, kept for long, at room temperature, in summer months are open invitations for such microbes to make the food poisonous. In contrast to salmonellosis or butulism the attack does not last more than one day and usually wanes within 6 hours or less.

NATURAL ORGANIC TOXICANTS PRESENT IN SOME OF THE FOODS

(i) Solanine

It is a common name used collectively for a group of tozic glycoalkaloids found in a number of plants. Potatoes damaged or black-spotted potatoes may contain upto 400-500 ppm of this toxicant. Consumption of such potatoes may cause abdominal pain and diarrhoea.

(ii) Gosypoll

It is a toxicant made up of phenolic compounds and may be found in free state as well in combination with protein, in cotton seed and its flour, but never in cotton seed oil. It is toxic to man, pig sheep, chicken and fish, but not to the ruminants (cows and buffaloes). The maximum allowable concentration of this toxicant is 1.1 per cent in cotton seed fleur used for making biscuits.

(iii) Oxalic Acid

It is contained in chana dal (180 mg) piyas (407 mg), amla (296 mg) and khesari oil (lathyrus 122 mg) per 100 g. Consumption of heavy dose of this food produces acid cramps, haemohiliae and even death. But such poisoning is generally rare. However, it is advisable to restrict the intake of foods containing more than 0.5 per cent of the acid. Incidence of formation of stones in kidney is high among the poor population of Rajasthan, presumably because of their habitual heavy consumption of the above mentioned food items.

(iv) Erucic Acid

It is a long chain unsaturated fatty-acid normally present in mustard oil mixed with rape-seed oil in concentration exceeding to about 45 per cent. Habitual use of such mixed oil has been found to cause myocardial fibrosis (reduction in eye sight) in human being.

Noise Pollution

Noise pollution is often misinterpreted by most of the people as sound pollution. As a matter of fact, sound is a pure tone, harmonically related, having fixed frequencies and amplitudes, occurring at regular intervals and produces meaningful communication and pleasure in hearing. Musical lectures given by teachers, leaders and the preaching by saints are the examples of sound. On the other hand, 'unwanted sound' is termed as noise, having a complex mixture of a number of pure tones of various frequencies and amplitudes. These sound waves fluctuate and repeat themselves in highly haphazard manner. Noise pollution has become a part of industrial life in modern times Urban people are at constant exposure to such pollution contributed by disorderedly maintained vehicles Rural population are increasingly being exposed to noise pollution by 'loudspeakers', an amplifier of music especially during social functions. The people working in non industrial establishments are also vulnerable to noise pollution, significantly contributed by ill-maintained type-writers and electronic instruments.

Units of Noise Pollution Measurement

The unit for the expression of the Noise pollution intensity is decibel (db) *i.e.* the logarithmic ratio of noise intensity (I_0) to reference level (I). The minimum intensity of sound audible by man is termed as reference level (I), approximately equal to the sound pressure of 0.0002 microbars (dynes per cm² of an energy of about 10^{-16} watts). Mathematically the db is expressed as ;

$$db = 10 \log_{10} \frac{10}{1}$$

Accordingly, 10, 20 and 100 db represent 10, 10^2 and 10^{10} times the threshold intensity respectively. The threshold intensity of sound required for human hearing is zero db, the rattling of leaves produces 10 db, and sounds generated during broadcasting is 20 db. Sound level of 30 db is generated is bed room at night pillow talk. 40 db in library and 60 db in normal conventional speeches. The play of the radio at average volume induces a sound level of 70 db, light traffic produces noise at 74 db and 80 db of noise is produced by a typical factory. Running train generates a noise of 90 db, orchestra of 100 db, rock band of 110 db while an aircraft take off produces noise of 120 db. Threshold of pain is of 140 db.

Measurement of Noise Pollution

The instruments needed for the measurement of noise pollution are many and varied, depending upon the types of experiments to be done. For examples, sound level meter, noise dosimeter and frequency analyser are the instruments, commonly used for measuring the total intensity of noise, the actual noise that the worker receives during working hours and the frequency bands respectively. Brief description of each of the instruments is given below :

(i) Sound Level Meter

A sound level meter consists of a microphone, amplifier, rectifier and a display meter. The sound pressure level in db is directly read out from this instrument.

(ii) Noise Dosimeter

The different noise levels, existing close to or away from the machines are measured directly by this instrument. The total levels of noise that wearers receive during their working periods are indicated whereas the total intensity of sound produced in the campus are shown by the sound level meter.

(iii) Frequency Analyser

The frequency-cum-noise level distribution is measured by such type of instrument which helps to devise any control measures to reduce the potentially effective part of the noise level.

Acceptable Noise Standard

Sound is essential for life. Organisms are at constant exposure to some other forms of energy, without which the brains may not function. At the same time the adage, "Too much of every thing is bad", is also applicable in this case. The 'unwanted sound', what is called as noise, is not at all ecologically desirable at any level. However, the permissible noise levels in db is dependent upon the time of exposure, 80 db of noisc level for 8 hours days, 90 db of noise for 4 hours per day, 95 db of noise for 2 hours per day, 100 db of noise for 1 hour/day 105 db of noise for 1/2 hour per day, 110 db of noise for 1/4 hour (15 minute)/day and 115 db of noise for 1/8 hour per day have been permitted by almost all the countries of the world.

Exposures to continuous or intermittent noise level higher than 115 db should not be permitted. However, 90 db of noise level for 8 hours/day of exposure has been granted in case of mining occupations for workers in India. The maximum desirable limit for noise levels in residential area and recreational activities centres around the habitations has been fixed as 40 db in night, 45 db in evening and 50 db during day time.

Effect of Noise Pollution on Human Health

Noise level above 80 db for more than 8 hours a day has been found to increase tension in a human being of normal mental make-up. Noise level above 80 db during intermittent exposure produces a change in breathing pattern. Continued exposure to high levels of noise results in annoyance, fatigue, shifting of threshold limit of hearing and the worst may be the total loss of hearing. The bad effects of noise pollution have been seen to interrupt the process of blood circulation, to increase breathing amplitude to bring changes in size of the pupils of the eye and in the rate of saliva and gastric secretions. Noise pollution above 120 db produced by air craft and turbine have been shown to induce many adverse biochemical changes in a human being for 3.88 hours of exposure per day for a period of one to 6 months. The concentration of cholesterol in blood becomes elevated alongwith increased white cell counts in peoples living around close vicinities of busy aerodromes. Industrial workers, exposed to noise level of more than 100 db, reveal the tendency of reduced intra-ventricular conductivity besides hypertension.

Remedies of Noise Pollution

(i) Decibel meters should be installed along highways and in places of public gatherings such as picture halls, libraries, offices and hostels. Each industrial establishment must have such facilities in order to have a check on the intensity of noise pollution, being produced throughout the working period.

(ii) Greenbelt vegetation and open space, at regular intervals, within the densely populated areas of big towns, must be developed. A dense evergreen hedge can reduce the noise by 10 db during garbage collection in the town. Border planting along highways can very effectively control the noise pollution if plantings be done in lower side with respect to the source of noise production and be on high side towards the listener.

(iii) A strip of 20 feet wide plantation inside the compound wall, effectiveiy protects the houses from sound pollution, contributed by vehicular traffic.

(iv) Technological modification is suggested to be achieved so that a sound associated with minimum possible frequency be produced from the machines.

14

The Environment Protection Act, 1986

No. 29 of 1986

An Act to provide for the protection and improvement of environment and for matters connected therewith.

Whereas decisions were taken at the United Nations Conference on the Human Environment held at Stokholm in June, 1972, in which India participated, to take appropriate steps for the protection and improvement of human environment.

And whereas it is considered necessary further to implement the decisions aforesaid in so far as they relate to the protection and improvement of environment and the prevention of hazards to human beings, other living creatures, plants and property.

Be it enacted by Parliament in the Thirty-seventh Year of the Republic of India as follows :

CHAPTER I
Preliminary

Short Title, Extent and Commencement

1. (1) This Act may be called the Environment (Protection) Act, 1986.

(2) It extends to the whole of India.

(3) It shall come into force on such date as the Central Government may, by notification in the Official Gazette, appoint and different dates may be appointed for different provisions of this Act and for different areas.

Definitions

2. In this Act, unless the context otherwise requires.

(a) "Environment" includes water, air and land and the inter-relationship which exists among and between water, air and land, and human beings, other living creatures, plants, micro-organism and property.

(b) "Environmental pollutant" means any solid, liquid or gaseous substance present in such concentration as may be, or tend to be, injurious to environment.

(c) "Environmental pollution" means the presence in the environment of any environmental pollutant.

(d) 'Handling", in relation to any substance, means the manufacture, processing, treatment, package, storage, transportation, use, collection, destruction, conversion, offering for sale, transfer or the like of such substance.

(e) "Hazardous substance" means any substance or preparation which, by reason of its chemical or physio-chemical properties or handling is liable to cause harm to human beings, other living creatures, plants, micro-organism, proserty or the environment.

(f) "Occupier", in relation to any factory or premises, means a person who has control over the affairs of the factory or the premises and includes, in relation to any substance, the person in possession of the substance.

(g) "Prescribed" means prescribed by rules made under this Act.

CHAPTER II

General Powers of the Central Government

Power of Central Government to take measures to protect and improve environment

3. (1) Subject to the provisions of this Act, the Central Government shall have the power to take all such measures as it deems necessary or expedient for the purpose of protecting and improving the quality of the environment and preventing, control-over and abating environmental pollution.

(2) In particular and without prejudice to the generality of the provisions of sub-section (1), such measure may include measure with respect to all or any of the following matters, namely :

(i) Co ordination of actions by the State Governments, officers and other authorities :

(a) Under this Act, or the rules made thereunder; or

(b) Under any other law for the time being in force which is relatable to the objects of this Act;

(ii) Planning and execution of a nationwide programme for the prevention, control and abatement of environmental pollution;

(iii) Laying down standards for the quality of environment in its varions aspects;

(iv) Laying down standards for emission or discharge of environmental pollutants from various sources whatsoever.

Provided the different standards for emission or discharge may be laid down under this clause from different sources having regard to the quality or composition of the emission or discharge of environmental pollutants from such sources.

(v) Restriction of areas in which any industries, operations or processes or class of industries, operations of processes shall not be carried out or shall be carried out subject to certain safeguards.

(vi) Laying down procedures and safeguards for the prevention of accidents which may cause environmental pollution and remedial measures for accidents.

(vii) Laying down procedures and safeguards for the handling of hazardous substances.

(viii) Examination of such manufacturing processes, materials and substances as are likely to cause environmental pollution.

(ix) Carrying out and sponsoring investigations and research relating to problems of environmental pollution.

(x) Inspection of any premises, plant, equipment, machinery, manufacturing or other processes, materials or substances and giving, by order, of such directions to such authorities, officers or persons as it many consider necessary to take steps for the prevention, control and abatement of environmental pollution.

(xi) Establishment or recognition of environmental laboratories and institutes to carry out the functions entrusted to such environmental laboratories and institutes under this Act.

(xii) Collection and dissemination of information in respect of matters relating to environmental pollution.

(xiii) Preparation of manuals, codes or guides relating to the prevention, control and abatement of environmental pollution.

(xiv) Such other matters as the Central Government deems necessary or expedient for the purpose of securing the effective implementation of the provisions of this Act.

(3) The Central Government may, if it considers necessary or expedient so to do for the purposes of this Act, by order, published in the Official Gazette, constitute an authority or authorities by such name or names as may be specified in the order for the purpose of exercising and performing such of the powers and functions (including the power to issue directions under Section 5) of the Central Government under this Act and for taking measures with respect to such of the matters referred to in Sub Section (2) as may be mentioned in the order and subject to the supervision and the control of the Central Government and provisions of such order such authority or authorities may exercise the powers of perform the functions or take the measures so mentioned in the order as if such authority or authorities had been empowered by this Act to exercise those powers or perform those functions or take such measures.

Appointment of Officers and their Powers and Functions

4. (1) Without prejudice to the provisions of Sub-section (3) of section 3, the Central Government may appoint officers with such designations as it thinks fit for the purposes of this Act and may entrust to them such of the powers and functions under this Act as it may deem fit.

(2) The officers appointed under sub-section (1) shall be subject to the general control and direction of the Central Government or, if so directed by the Government, also of the authority or authorities, if any, constitued under sub-section (3) or if any other authority or officer.

Power to give Directions

5. Notwithstanding anything contained in any other law but subject to the provisions of this Act, the Central Government

may, in the exercise of its powers and performance of its functions under this Act, issue directions in writing to any person, officer or any authority and such person, officer or authority shall be bound to comply with such directions.

Explanation : For the avoidance of doubts, it is hereby declared that the power to issue directions under this section includes the power to direct :

(a) The closure, prohibition or regulation of any industry, operation or process; or

(b) Stoppage or regulation of the supply of electricity or water or any other service.

Rules to Regulate Environmental Pollution

6. (1) The Central Government may, by notification in the Official Gazette, make rules in respect of all or any of the matters refegred to in section 3.

(2) In particular, and without prejudice to the generality of the foregoing power, such rules may provide for all or any of the following matters, namely :

(a) The standards of quality of air, water or soil for various areas and purposes.

(b) The maximum allowable limits of concentration of various environmental pollutants (including noise) for different areas.

(c) The procedures and safeguards for the handling of hazardous substances.

(d) The prohibition and restrictions on the handling of hazardous substances in different areas.

(e) The prohibition and restrictions on the location of industries and the carrying on processes and operation in different areas

(f) The procedures and safeguards for the prevention of accidents which may cause environmental pollution and for providing for remedial measures for such accidents.

CHAPTER III

Prevention, Control and Abatement of Environmental Pollution

7. No person carrying on any industry, operation or process shall discharge or emit or permit to be discharged or emitted any environmenral pollutant in excess of such standards as may be prescribed.

8. No person shall handle or cause to be handled any hazardous substance except in accordance with such procedure and after complying with such safeguards as may be prescribed.

9. (1) Where the discharge of any environmental pollutant in excess of the prescribed standards occurs or is apprehended to occur due to any accident or other unforeseen act or event, the person responsible for such discharge and the person in charge of the place at which discharge occurs or is apprehended to occur shall be bound to prevent or mitigate the environmental pollution caused as a result of such discharge and shall also forthwith :

(a) Intimate the fact of such occurrence or apprehension of such occurrence; and

(b) Be bound, if called upon, to render all assistance, to such authorities or agencies as may be prescribed.

(2) On receipt of information with respect to the fact or apprehension of any occurrence of the nature referred to in sub-section (1), whether through intimation under that sub-section or otherwise the authorities or agencies referred to in sub-section (i) shall, as early as practicable, cause such remedial measures to be taken as are necessary to prevent or mitigate the environmental pollution.

(3) The expenses, if any incurred by any authority or agency with respect to the remedial measures referred to in sub-section (2), together with interest (at such reasonable rate as the Government may, by order, fix) from the date when a demand for the expenses is made until it is paid, may be recovered by such authority or agency from the person concerned as arrears of land revenue or public demand.

Power to Entry and Inspection

10. (1) Subject to the provisions of this section, any person empowered by the Central Government in this behalf shall have a right to enter, at all reasonable times with such assistance as he considers necessary, any place :

(a) For the purposs of performing any of the functions of the Central Government entrusted to him;

(b) For the purpose of determining whether and if so in what manner, and such functions are to be performed or whether any provisions of this Act or the rules made thereunder or any notice, order, direction or authorisation served, made, given or granted under this Act is being or has been complied with;

(c) For the purpose of examining and testing any equipment, industrial plant, record, register, document or any other material object or for conducting a search of any building in which he has reason to believe that an offence under this Act or the rules made thereunder has been or is being or is about to be committed and for seizing any such equipment, industrial plant, secord, register, document or other material object if he has reasons to believe that it may furnish evidence of the commission of an offence punishable under this Act or the rules made thereunder or that such seizure is necessary to prevent or mitigate environmental pollution.

(2) Every person carrying on any industry, operation or process or handling any hazardous substance shall be bound to render all assistance to the person empowered by the Central Government under sub-section (1) for carrying out the functions under that sub-section and if he fails to do so without any reasonable cause or excuse, he shall be guilty of an offence under this Act.

(3) If any person will fully delays or obstructs any person empowered by the Central Government under sub-section (1) in the performance of his functions, he shall be guilty of an offence under this Act.

(4) The provisions of the Code of Criminal Procedure, 1973, or, in relation to the State of Jammu and Kashmir, or any area in which that Code is not in force, the provisions of any corresponding law in force in that State or area shall, so far as may be, apply to any search or seizure under this section as the they apply to any search or seizure made under the authority of a warrant issued under section 94 of the said Code or, as the case may be, under the corresponding provisions of the said law.

Power to Take Sample and Procedure to be Followed in Connection Therewith

11. (1) The Central Government or any officer empowered by it in this behalf, shall have power to take, for the purpose of analysis, samples of air, water, soil or other substance from any factory, premises or other place in such manner as may be prescribed.

(2) The result of any analysis of a sample taken under sub-section (1) shall not be admissible in evidence in any legal proceeding unless the provisions of sub-sections (3) and (4) are complied with.

(3) Subjected to the provisions of sub section (4), the person taking the sample under sub-section (1) shall :

(a) Serve on the occupier or his agent or person in charge of the place, a notice, then and there, in such form as may be prescribed, of his intention to have it so analysed;

(b) In the presence of the occupier or his agent or person, collect a sample for analysis;

(c) Cause the sample to be placed in container or containers which shall be marked and sealed and shall also be signed both by the person taking the sample and occupier or his agent or person;

(d) Send without dealy, the container or the containers to the laboratory established or recognised by the Central Government under section 12.

(4) When a sample is taken for analysis under sub-section (1) and person taking the sample serves on the occupier or his agent or person, a notice under clause (a) of sub-section (2), then :

(a) In a case where the occupier, his agent or person wilfully presents himself, the person taking the sample for analysis to be placed in a container or containers which shall be marked and sealed and shall also be singned by the person taking the sample, and

(b) In a case where the occupier or this gent or person present at the time of taking the sample refuses to sign the marked and sealed container or containers of the sample as required under clause (c) of sub-section (3), the marked and sealed container or containers shall be signed by the person taking the samples, and the container or containers shall be sent without delay by the person taking the sample for analysis to be laboratory established or reoognised under section 12 and such person shall inform the Government Analyst appointed or recognised under section 13 in writing, about the wilful absence of the occupier or his agent or person, as the case may be, his refusal to sign the container or containers.

Environmental Laboratories

12. (1) The Central Government may, by notification in the Official Gazette :

(a) Establish one or more environmental laboratories;

(b) Recognise one or more laboratories or institutes as environmental laboratories to carry out the functions entrusted to an environmental laboratory under this Act.

(2) The Central Government may, by notification in the Official Gazette, make rule specifying :

(a) The functions of the environmental laboratory;

(b) Procedure for the submission to the said laboratory of samples of air, water, soil or other substance for analysis or tests, the form of the laboratory report thereon and the fees payable for such report;

(c) Such other matter as may be necessary or expedient to enable that laboratory to carry out its functions.

Government Analyst

13. The Central Government may by notification in the Official Gazette, appoint or recognise such persons as it thinks fit and having the prescribed qualifications to be Government Analyst for the purpose of analysis of samples of air, water, soil or other substance sent for analysis to any environmental laboratory established or recognised under sub-section (1) of section 12.

Repor of Government Analyst

14. Any document purporting to be a report signed by a Government analyst may be used as evidence of the facts stated therein in any proceeding under this Act.

Penalty for Contravention of the Provisions of the Act and the Rules, Orders and Directions

15. (1) Whoever fails to comply with or contravenes any of the provisions of this Act, or the rules made or orders or directions issued thereunder, shall, in respect of each such failure or contravention, be punishable with imprisonment for a term which may extend to five years or with fine, which may extend to one lakh ruppes, or with both, and in case of the failure or contravention continues, with additional fine which may extend to five thousand rupees for every day during which such failure or contravention continues after the conviction for the first such failure or contravention.

(2) If the failure or contravention referred to in sub-section (1) continues beyond a period of one year after the date of conviction, the offender shall be punishable with imprisonment for a term which may extend to seven years.

Offences by Companies

16. (1) Where any offence under this Act has been committed by a company, every person who, at the time the offence was committed, was directly in charge of, and was responsible to the company for the conduct of the business of the company; as well as the company, shall be deemed to be guilty of the offence and shall be liable to be proceeded against and punished accordingly :

Provided that nothing contained in this sub-section shall render any such person liable to any punishment provided in this Act, if he proves that the offence was committed without his knowledge or that he exercised all due diligence to prevent the commission of such offence.

(2) Notwithstanding anything contained in sub-section (1), where an offence under this Act has been committed by a Department of Government and it is proved that

the offence has been committed with the concert and connivance of, or is attributable to any neglect on the part of, any director, manager, secretary or other officer of the company, such director, manager, secretary or other officer shall also be deemed to be guilty of that offence and shall be liable to be proceeded against and punished accordingly.

Explanation : For the purposes of this section.

(a) "Company" means any body corporate and includes a firm or other association of individuals;

(b) "Director", in relation to a firm, means a partner in the firm.

Offences by Government Departments

17. (1) Where an offence under this Act has been committed by any Department of Government, the Head of the Department shall be deemed to be guilty of the offence and shall be deemed to be liable to be proceeded against and punished accordingly.

Provided that nothing contained in this section shall render such Head of the Department liable to any punishment if he proves that the offence was committed without his knowledge or that he exercised all due diligence to prevent the commission of such offence.

(2) Notwithstanding anything contained in sub-section (1), where an offence under this Act has been committed by a Department of Government and it is proved that the offence has been committed with the consent or connivance of, or is attributable to any neglect on the part of any officer, other than the Head of the Department, such officer shall also be deemed to be guilty of that offence and shall be liable to be proceeded against and punished accordingly.

CHAPTER IV
Miscellaneous

Protection of Action Taken in Good Faith

18. No suit prosecution or other legal proceeding shall be taken against the Government or any officer or other employee of the Government or any authority constituted under this Act or any member, officer or other employee of such authority in respect of anything which is done or intended to be done in good faith in pursuance of this Act or the rules made or orders of directions issued thereunder,

Cognizance of Offence

19. No courts shall take congnizance of any offence under this Act except on a complaint made by :

(a) The Central Government or any authority or officer authorised in this behalf by that Government; or

(b) Any person who has given notice of not less than sixty days, in the manner prescribed, of the alleged offence and of his intention to make a complaint, to the Central Government or the authority or officer authorised as aforesaid.

Information, Reports or Returns

20. The Central Government may, in relation to its functions under this Act, from time to time, require any person, officer, State Government or other authority to furnish to it or any prescribed authority to furnish to it or any prescribed authority to furnish to it or any prescribed authority or officer any reports, returns, statistics, accounts and other information and such person, officer, State Government or other authority shall be bound to do so.

Members, Officers and Employees of the Authority Constituted Under Section 3 to be Public Servants

21. All the members of the authority, constituted, if any, under section 3 and all officers and other employees of such

authority when acting or purporting to act in pursuance of any provisions of this Act or the rules made or orders or directions issued thereunder shall be deemed to be bublic servents within the meaning of section 21 of the Indian Penal Code.

Bar of Jurisdiction

22. No civil court shall have the jurisdiction to entertain any suit or proceeding in respect of anything done, action taken or order or direction issued by the Central Government or any other authority or officer in pursance of any power confered by or in relation to its functions under this Act.

Power of Delegate

23. Without prejudice to the provisions of sub-section (3), of section 3, the Central Government may, by notification in the Official Gazette, delegate, subject to such conditions and limitations as may be specified in the notification, such of its powers and functions under this Act (except the power to constitute an authority under sub-section (3) of section 3 and to make rules under section 25) as it may deem necessary or expedient, to any cfficer, State Government or other authority.

Effect of Other Laws

24. (1) Subject to the provisions of sub-section (2), the provisions of this Act and the rules or orders made therein shall have effect notwithstanding anything inconsistent therewith contained in any enactment other than this Act.

(2) Where any act or omission constitutes an offences punishable under this Act and also under any other Act then the offender found guilty of such offence shall be liabie to be punished under the other Act and not under this Act.

Power to Make Rule

25. (1) The Central Government may, be notification in the Official Gazette, make rules for carrying out the purposes of this Act.

(2) In particular, and without prejudice to the generality of the foregoing power, such rules may provide for all or any of the following matters, namely :

(a) The standards in excess of which environmental pollutants shall not be discharged or emitted under section 7;

(b) The procedure in accordance with and the safeguards in compliance with which hazardous substances shall be handled or caused to be handled under section 8;

(c) The authorities or agencies to which intimation of the fact of occurence or apprehension of occurrence of the discharge of any environmental pollutant in excess of the prescribed standards shall be given and to whom all assistance shall be bound to be rendered under sub-section (1) or section 9;

(d) The manner in which samples of air, water, soil or other substance for the purpose of analysys shall be taken under sub-section (1) of section 11;

(e) The from in which notice of intention to have a sample analysed shall be served under clause (a) sub-section (3) of section 11;

(f) The functions of the environmental laboratories, the procedure for the submission to such laboratories of samples of air, water, soil and other substances for analysis or test; the form of laboratory report; the fees payable for such report and other matters to enable such laboratories to carry out their functions under sub-section (2) of section 12;

(g) The qualifications of Government Analyst appointed or recognised for the purpose of analysis of samples of air, water, soil or other substances under section 13;

(h) The manner in which notice of the offence and of the intention to make a complaint to the Central Government shall be given under clause (b) of section 19;

(i) The authority of officer to whom any reports, returns, statistics, accounts and other information shall be furnished under section 20;

(j) Any other matter which is required to be, or may be, prescribed.

Rules Made Under this Act to be Laid Before Parliament

26. Every rule made under this Act shall be laid, as soon as may be after it is made, before each House of Parliament, while it is in session, for a total period of thirty days which may be comprised in on session or in two or more successive sessions, and if, before the expiry of the session immediatley following the session or the successive sessions afforesaid, both Houses agree in making any modification in the rule or both Houses agree that the rule should not be made, the rule shall thereafter have effect, as the case may be; so however, that any such modification or annulment shall be without prejudice to the validity of anything previously done under that rule.

15

Environmental Education

Introduction

Nature is fascinating. It embodies the spirit of the creator. The soft green leaf moving gently in the breeze, the drop of dew quivering on the petal of a rose, the koel singing hidden in a mango grove, all are so enchanting, so enthralling. No less is the of hild with its hazel eyes, silken, hair, rosy cheeks, tooth less mouth and tender skin. Its father, the man, too is very handsome and is perhaps the most unique creation of God, unique in the sense that it is man who has made the world so beautiful. But for him, nature would not have been so enjoyable as it is. He has converted wild forests into parks, gardens, orchards and cities, and deserves all praise for that. But in doing that an in his attempts to make life more and more comfortable, he often destroyed forests thoughtlessly, poluted air and water recklessly, despoiled nature ruthlessly. The inevitable consequence was an ominious deterioration of the environment. He did not stop even then. Nor did he use that much caution and took those precautions which should have been taken He, in fact, jumped headlong towards environmental pollution and crisis.

Now the time has come when he should be careful. If he does not collect himself and sobser down, he is bound to find himself in an irreversible predicament, in an ecological catastrophe which will bring an end of the Homo sapiens on this lovely planet of ours. How to avert such a catastrophe? (Obviously environmental education is the answer). Environmental education will not only open our eyes to the havoc we have wrought on ourselves in the form of by-products of the progress attained by us but also it will inform us as to how we can intertwine progress and life so that they exist together. We have, therefore, to convince ourselves that it is high time we realised the need of environmental education. Had we done that earlier, we could have saved some species of animals from being wiped away. Needless to mention, the rate of extinction of species is disappearing every year, we can save these species only if we organise environmental education programmes on a large scale.

Environmental education is education through, about and for environment. Its scope is, therefore, very wide. It begins from using environment as a medium of learning, and includes all that Kalidas, Wordsworth and others have said in appreciation of nature and also all that scientists and schoolars have disclosed about our physical and social environment, and finally it includes all that we say and do for conserving our resources and for beautifying our surroundings including urban and country planning. Thus, it will be seen that even before the formation of the National Association on Environmental Education in the U.K. in 1972 and the enactment of the law on environmental protection and conservation in 1972 in the US.A., some sort of environmental education was being carried out by poets, naturalists, ornithologists, town and country planners, etc. The emphasis has, however, recently been shifted from individuals and specific groups to nations and also to international bodies such as Unesco and International Union for the Conservation of Nature in the hope that it may help people how hazardous environmental pollution, population explosion and resource depletion could be.

Meaning and Scope

But for a few dens and dungeons made by man, his environment is so beautiful and lustrous; the azure sky, the blue lakes, the starry nights, the rolling waves, the crescent moon, the conifers, the jacarandas, the gulmohars, the shade trees, the flowers the insects, the breeze, the brooks, the birds, the wild life, etc. all are breath-taking. Wordsworth was rightly drawn towards them. His heart lead, up when he beheld a rainbow in the sky, or when daffodils flashed upon that inward eye which was a bliss of solitude for him. Not only the natural environment but also the Taj Mahal, the Qutub Minar and many other historical, social and cultural places fascinate men day in and day out. Why shall they be not used in their natural setting as a teaching-learing aid ? Why not teach about a brook or a bird or a butterfly when children are attracted by it, instead of teaching the topic within the leaden walls of the classroom ? In short, a lot of teaching-learing can be carried out through environment. And that is the first aspect of environmental education.

Its second aspect covers teaching-learning about environment. That is important, for man has to tackle his environment every day for his survival, sustenance and prosperity. He cannot escape it under any circumstances, right from birth to death. He is born in a family and is brought up by the family which is a primary group for him, when he grows up beyond his infancy and becomes a child, and after that a teenager, then an adult, he moves about in his neighbourhood and in the community and participates in the activities of many secondary and tertiary groups. It is, therefore, in one's own interest that one should learn about environment, more so the immediate one; one should know the trees in one's neighbourhood, how they area planted, how they are used for fuel and energy, for fruit and food and foar the beautification of one's surrounderings. One has also to know which trees and crops can grows in one's locality. This, in turn, inbolves a knowledge of the chemistry, texture and structure of the soil of the locality. It also involves a study of the drainage system, of the rocks and of the running water, for many plants cannot grow in water-logged soil or on ussar l nd. May be, one's house lies in a drainage basin into which rain water flows through some channels or from some slopes and flows out of the basin through some streams or rivulets which subsequently join a river, or perhaps there may be some place where it gets deposited and created a waterlogging problem. Mosquitoes may breed theare and cause malaria. For similar reasons, it is very necessary to know about several other features of natural environment which affect man's food supply, health and recreation.

Equally necessary it is to know about the social institutions in the community such as the schools, the banks, the municipality, the panchayat or other agencies of local self-government, the items on which taxes are levied, the types of taxes, the items of public expenditure, the source from which the community gets its electricity supply, the authority responsible for building and maintaining roads, local industries, the religions practised, within the community, local firms where people can sek employment, skill and education to be possessed by them, the rate of population growth in the community, etc. Without a knowledge of these, we cannot lead a satisfying life. Without knowing about the agencies responsible for maintaining law and order and those responsible for civic life, and without discharging our duties and rights as responsible citizens, we would neither be promoting our own good nor that of the society.

Besides the foregoing reasons, one other which highlights the need for learing about environment is that such learning is both rewarding and enjoyable and it is for that reason that people have always been sallying out to far off destinations. Marcopolo, Hue'n Tsangh, Ibn Batuta, Columbus and many others undertook hazardous journeys to distant lands to know people and places. Even todauy we undertake strenuous journeys to unknown places within the country or overseas just for fun and pleasure or just to quench our curiosity about other cultures and countries; many of the airports of the world are througed with groups of travellers who trot round the globe to satisfy their wander lust. Those who cannot afford such extensive tours, travel within their state. Those who cannot afford even that go out whenever they are able to manage funds and time to nearby fairs or temples.

Perhaps man is an animal and he still shares with other animals his wander lust. He may be living in a cosy niche in a multi-storeyed building or in a bungalow on the bank of a stream or in a homestead in the middle of an agricultural farm, or in a bustee of an industrial metropolis, he wants to go out once in a while to see the grandeur of the hills, or to look the expanse of the sea, or to watch birds in a sanctuary, or to roam like Pan in a wood, or to go to the countryside to his ancestral house. He bears the expense and inconvenience involved in the travel, for like other animals he too belongs to an ecosystem and wants to get closer to it or may be he wants to get closer to the ecosystem in which his remote ancestors lived. He wants to brush aside leaves and twigs, to tread on grass, to watch other animals and to live at that moment like an animal free of all cares and worries. This shows that environment is a source of happiness to man and his happiness will be heightened, all the more, if he learns to love and appreciate nature. Not only that, since he is a social animal, he wants the company of other people. Hence the need to understand social environment.

A third aspect of environmental education is education for environment. Environment during the present century, to use Max Nicholson's phase, has undergone revolution : population has exploded, physical resources are being increasingly depleted and technological growth has been phenomenal, resulting in an unprecedented air, water and noise pollution, land dereliction and accumulation of pesticide residues. All these facts remind us of Thanatos, of death instinct, which may sweep the entire world. We have started to realise, and are rather haunted by the feeling that the world is moving headlong towards disaster, or perhaps we might have reached its brinks and may soon destroy ourselves. It is, therefore, very urgent that we may diagnose this malady and adopt preventive and curative measures.

In this connection the first problem that merits our attention is that of population explosion, the surge of teeming millions in the third world, the developing countries. With their meagre resources, and owing to the limit of this planter's population-bearing capacity, and for the reason that the growing masses are submerged neck deep in the mire of poverty we have to pay attention to this problem. Its enormity is further proved by the fact that as of date 700 million people out of the total lot of 2600 million living in the developing world are awfully poor and are breathing between life and death. Any moment starvation may shove them into the open jaws of death. Not only that, malaria, trachoma, bilharzia and tuberculosis afflict about 1000 million people. An equal number do not have safe drinking water. These problems are multiplying every day. Every day the pressure of population on agricultural land is increasing. And since this land cannot employ all these people, they have to march to towns to work as unskilled labourers, and in most of the cases to dwell on footpath, or to lie in a hole in some expanding shanty-town, or in a slum in some ghetto, soon to fall prey to some disease which their malnutritioned bodies are not able to resist, or soon to age and become old at the age of thirty-five. How sad ! And what is the remedy ? The inevitable answer is population control.

The second problem is that of the depletion of physical resources, which are of two types-renewable and non-renewable. The renewable ones are soil, water, forest, vegetation and wild life. The non-renewable ones are minerals, oil and coal. This classification, however, does not give us the license to use the renewable resources ruthlessly. If we cut forests or destroy our plants without caring for them, the equilibrium in nature would be disturbed. The food chain would be affected. Many animals which live on vegetation would die, and in turn, many carnivores which feed on these animals would die and in the process man will be affected, Not only this, even the humus and top layers of soil may be washed off by rain and running water causing soil erosion. Thus, erratic use and destruction of even renewable resources may land us into such serious ecological problems which may be difficult to be solved, and which may result in the wiping away of many species of animals, so necessary to maintain equilibrium in nature.

Similarly, reckless use of non-renewable physical resources may land us into serious difficulties. The depletion of fossil fuel, the exhaustion of petroleum supplies and the wearying out of other physical sources of energy may force us to use nuclear energy in a big way which may create new problems of environmental pollution. Probably, we may have to concentrate more on the commercial use of solar energy, most of which is being wasted owing to non-uti isation. Apart from these problems, there is also the problem of the exhaustion of ores such as rock phosphate, copper etc. which are non-renewable. Hence the main problem is how to use there resources prudently and save them sagaciously for future use.

The third problem is that of controlling environmental pollution which is increasing, by and by, on account of technological growth. This problem is more severe in western countries, but its existence cannot be denied in our country too. The pollution of the Ganges with the waste products of the factory at Rishikesh which has reached magnitude of its own is a case in point. A second example is that of the carbon monodioxide emisted by the chimneys of thermal power stations and other factories. Moreover, pollution is also caused by automobiles, buses and trucks These problems of pollution are further aggravated by noise pollution which too sometimes becomes menacing. The noise of traffic in the metropolises, the thundering sound of the supersonic jets, and the roaring of the mills, all pollute the atmosphere. We have to think of controlling this pollution too, as we are thinking of conrrolling air and water pollution, if ecological balance is to be maintained and if living being are to be saved from extinction.

In short, it can be said that environmental education is education through environment, about environment and for environment It is both a style and subject-matter of education. In so far as the style is concerned, it means using environment in so far as the teaching for environment is concerned, it means controlling the environment, as a teaching-learning aid and as an approach to education. In so far as the subject matter or content is concerned, it means teaching about the components and constituents of environment, establishing proper ecological equilibrium which entails proper use and conservation of resources and also involves control of environmental pollution. It also includes proper environmental planning so that environment is not only functionally useful but is also aesthetically enjoyable. This aspect would include horticultural planning, rural planning and urban planning.

This definition of environmental education appears on first reaction to be simplistic and superficial but actually it is comprehensive and deep, for it denotes that environmental education is a medium and process of education and that it covers man's relationship with his natural as well as social and man-made environment, and also it includes the relationship of population, industrialization, pollution, resource allocation and depletion, conservation, transportation, technology, energy and urban and rural planning to the total biosphere. Viewed so, the nexus of environmental education is multi-disciplinary in character and its quintessence is a commitment on the part of one and all, on the part of all us inhabiting this planter 'earth', to prevent deterioration of air, water, land and physical and social environment including inter-relations among people and among peoples so that a nuclear war, a chemical warefare or any other cataclysm generated by man may not destroy the world. Further, environmental education should be a lifelong process and should aim at not merely imparting knowledge and understanding of man's total environment and of the methods and their application for improving our near and surroundings but it should also aim at inculcating skills, attitudes and values necessary to understand, appreciate and improve our biosphere and troposphere.

It may be pertinent now to quot the definitions of environmental education formulated by varions agencies or organisations. The one contained in the United States Environmental Education Act, 1970 is as follows :

"For the purposes of this Act, the term 'Environmental Education means the educational process dealing with man's relationship with his natural and man-made surroundings, and includes the relation of population, pollution, resource allocation and depletion, conservation, transportion, technology, and urban and rural planning to the total human environment".

The First Report of the British Royal Commission on Environmental Pollution (1971) says :

"The best insurance for the environment is a commitment on behalf of the public to prevent the deterioration for air, water and land".

The International Union for the Conservation of Nature and Natural Resources Commission of Education has remarked in the International Working Meeting on Environmental Education in the School Curriculum, held under the auspices of Unesco in Paris in 1970 :

"Environmental education is the process of recognizing values and clarifying concepts in order to develop skills and attitudes necessary to understand and appreciate the inter-relateness among man, his culture and his biophysical surroundings. Environmental education also entails practice in decision-making and self-formulation of a code of behaviour about issues concerning environmental quality".

The Finnish National Commission in a seminar held in 1974 has said :

"Environmental education is a way of implementing the goals of environmental protection. Environmental education is not a separate branch of science or subject of study. It should be carried out according to the principle of lifelong integral education".

The National Anti-Pollution Law of Japan made in 1969 has stressed the following aspects of Environmental Education :

"In order to enable people to enjoy good health and a high quality of life, it is vital to prevent harmful effects to human health or damage to the environment caused by pollution of air, water and soil, noise, vibration, noxious smells, etc. caused by firms and individuals. The environment includes animals and plants and their ecological systems which are closely bound to the livelihood of people,

The Report of a Conference of African Educators. EDC and CREDO held at Nairobi in 1968 days :

To create an awareness and an understanding of the evolving social and physical environment as a whole, its natural, man-made, cultural, spiritual resource, together with the rational use and conservation of these resources for development."

The Aforesaid quotations would remain incomplete unless the definition formulated in the Tbilsi conference which was held at Georgia in the USSUR in October 1977 under the auspices of the Unesco and UNEP (United Nations Environment Programme) is also reproduced. The Tbisi Conference resolved that environmen al education should :

- consider the environment in its totality-natural and built, technological and social (economic, political, technological, cultural, historical, moral, aesthetic);
- be a continuous, lifelong process, beginning at the pre-school level and continuing through all formal and non-formal stages;
- be inter-disciplinary in its approach, drawing on the specific content of each discipline in making possible a holistic and balanced perspective;
- examine major environmental issues from the local, national, regional and international points of view so that students receive insight into environmental conditions in other geographical areas;
- focus on current and potential environmental situations while taking into account the historical perspectives;
- promote the value and necessity of local, national and international cooperation in the prevention and solution of environmental problems;
- explicity consider environmental aspects in plans for development and growth;

- enable learners to have a role in planning their learning experiences and provide an opportunity for making decisions and accepting their consequences;
- relate environmental sensitivity, knowledge, problem-solving skills and values-clarification to every ge, but with special emphasis on environmental sensitivity to the learner's own community in early years;
- help learners discover the symptoms and real causes of environmental problems;
- emphasize the complexity of environmental problems and thus the need to develop critical thinking and problem-solving skills;
- utilize diverse learning environments and a broad array of educational approaches to teaching/learning about and from the environment with due stress on practical activities and first-hand experience.

Background and Principles

Environmental education tends to undo the imphasis on rote-learning being laid even now as an inheritance of the medieval system of education.

During the middle ages when the societies were governed by Papal or Brahmanical or other priestly authorities, the maxim 'Believe and do not ask questions' was followed in schools. This religious despotism of the society coloured the functioning of the schools where students were ruled with rods and ferules. In fact, the philosophy that prevailed then was that a chid's mind was a receptacle in which teachers with the force of their indisputable authority could fill in any knowledge. Accordingly, instruction in schools meant memorisation of facts which teachers taught their pupils, that, it involved a lot of rote-learning which merely meant repeating the words stotras and shlokas or Ayats etc. Correctly without bothering about their meanings. Thus, the very spirit of learning, which is understanding, was sacrificed to the letter, that is, the cramming of facts. A concomitant of this rote-learning was rule-teaching which meant teaching rules before teaching their basis and background. Accordingly grammer and syntax, which are a collection of rules, were taught before language and prosody before poetry and multiplication tables before fundamental mathematical operations.

This sort of teaching was contrary to the fundamental nature of the child, for it involved learning inferences, abstractions and conceptions without an acquaintance with concrete and tangible properties of things. Under it the learner felt the burden of studies as though it were a heavy nioad which he was carrying and which he would be glad to throw off any moment. Moreover, knowledge acquired without using senses, and conceptions formed without correcting preceptions were often erroneous, falacious and misleading, for illustration, earth was considered flat before it was perceived to be round on the basis of sense data. Wise people were not satisfied with these conceptions. Even so, they could not protest much, for perhaps they feared persecution, may be meeting the same fate which whipped copernicus and swallowed Galelio. Gradually, however, favourable winds which aimed at changing the very tone and tenor of the society started blowing. Luther and Calvin presaged a new era which was ushered in by Kousseau. Bacon and others in the west and by Raja Ram Mohan Roy, Lokmanya Tilak, Dayanand Saraswati, Vivekananda, etc. in. ia.

Again the social change so brought was mirrored in the classroom and the new era did not only mark the freedom of man in the adult world but also that of his child in school. And the tyrannical and atrocious atmosphere of school described so painstakingly by Charles Dicken, started getting relaxed and was welcome to pupils. In sequel, the old method of presenting truths in abstract from started falling out of use and correspondingly a new method of presenting them in concrete form was adopted. This method involved the use of senses for acquiring learning experiences. Thus abacus, a ball-frame, was devised to be used by children for learning lessons in Arithmetic and a foot-rule for measuring heights and distances, and the measures, seers and pounds, for acquiring correct concept of weights. Geography came to be taught by using maps and globe. In brief, the students were encouraged to draw the truths or number, of form, and of relationship in position by observing objects presented to them in the classroom. This was certainly a positive step but it resulted sometimes in many contrived and artificil sort of models and teaching aids.

In addition to the caution and prudence to be observed in selecting proper teaching aids, what is perhaps more important is to use them and also to use the environment as a teaching-learning aid for developing in the students the systematic culture of the power of observation. That can be done only when the spontaneous activity of the child which always involves some or the otherkind of observation is not thwarted. In other words, childs spontaneous activity should not be construed by adults and teachers as purposeless action or play or mischief; it should rather be looked upon as a process of acquiring that knowledge on which all knowledge to be used, in adult life is to be used for instance, no one can claim to be a good physician unless that one has learnt observational skills, and it is owing to the use of these skills, that he is able to observe the symptoms of the disease in a patient and to make correct diagnosis. The same is true of an engineer about his work. In fact, observation of relationships of things which laymen overlook is the job of the specialist in relation to his own area. And this habit of keen observation cannot be developed in a short time. It is long drawn process, for skills develop gradually, and they develop most when one starts developing them since childhood

One of the methods of developing observational skills is encouraging children to explore, experience and assess their own environment and to modify it as far as possible to suit their own needs and the needs of their community and of the society at large. Keeping these objectives in view, emphasis is being increasingly laid on environmental education as an integral part of the school curriculam. The educational principles that buttress support to the inclusion of environmental education in school curriculum are given below :

1. Environmental education helps in programming learing experiences from simple to complex. For illustration, children look at a bird, observe its colourful plumage. see it eating figs or insects or watch it flying, etc. All this gives them some awareness of the eating and flying habits of birds, sometimes they mime their flying action and if they are not thwareted and are rather encouraged to express what they observed about birds, they will proceed from the aforesaid simple facts to learning many complex facts about birds, they will proceed from the aforesaid simple facts to learning many complex facts about birds; they learn about the structure of their feathers, about barbs and barbicules, about the difference between their contour and down feathers, about the weightlessness of the feather and the meaning of the idioms 'light as a feather' etc. Similarly, they can observe other phenomena, stars, moon, rainbow etc., and learn many facts, proceeding again from the simple to the complex ones. A corollary of this procedure would be that children would begin with few facts about one subject but would successively go on adding to facts and will finally learn many others linked with one another. Thus, the principle simple to complex, does not embrace only one but the entire gamut of inter-related facts from all subjects or an ensemble of knowledge. It may be added here that it is this principle that makes environmental education as a medium of learning different subjects.

2. Environmental education helps children to proceed from indefinite ideas to definite ones. In this connection, it may be said that the first perceptions and

thoughts of children are as vague as their first movement and the first attempt at speech, in fact, an infant is not able to distinguish between colours, but as it grows, it learns to discriminated among different and shades, environmental education helps in sharpening the development of these observational skills and hastens the transition of ideals in children's mind from indefiniteness to definitencess. For example, the child before seeing hills may have a vague idea about a valley or a dale or about the difference between a ravine and a gorge, or between a gorge and a glen or a canyon, but after seeing hills, becomes definite about them. He may also become definite about different sounds as that of the murmur of a brook or the coo-coo of a cuckoo. Similarly, there are many other ideas about which he may shed off his indefiniteness and may become definite. Thus, environment helps children to correct their perceptions and to proceed from indefiniteness of ideas to their definiteness.

3. Environmental education helps children to proceed from the concrete to the abstract. This is a very simple educational maxim and does not need any elaboration. Still it may be said that environment is full of concrete things which children may examine and classify and interpret and then draw their own conclusions and inferences about them. For example, children may observe different types of plants and animals and classify them according to their species genus family and order. They may be surprised to know that sugarcane and grass belong to the same botanical family, and that a shrew which weighs about 3 grams and a whale which weighs over 12,000 kilograms are both mammals, similarly they may establish many other relation ships which may appear queer to a layman, but are important to a scientist.

4. Environmental education helps the ordering of learning experiences from the empirical to the national. Needless to say, 'empirical to rational' is a very important educational maxim and like other educational maxims described above, was given the pride of place by no less an educationist than Herbert Spencer. This maxim is satisfied very well in environmental studies as children can observe phenomenon, and touch, feel and test it empirically and then draw their own conclusions and rational explanations. They may inspect, for instance, a number of banyan and neem trees and then say that the leaves of the former are simple and of the latter compound and that both have chlorophyll, and then after further exploration and observation conclude that plants form food in the presence of sunlight and then argue that sunlight is essential for plants 'survival'. Having argued so, they may verify their hypothesis experimentally.

5. A corollary of the foregoing principle which is so dear to the hearts of educationists is that education should help the child in the process of self-development. This means that children should be encouraged to conduct their own investigations and draw their own conclusions. They should be told as little as possible and made to discover as much as possible. This principle is very much consistent with what thinking men always do in life, namely, self-instruction which has been the wrap and woof of human progress. Now developing the habit of self-instruction and discovering things for themselves is possible maximally in programmes of environmental education. Children on exploring, observing and assessing the phenomenon around them not only engage in self instruction but also sometimes feel, they have discovered somethings new, and thus have the joy of discovery which is a source of great happiness for them.

6. Besides the forgoing principles, one other which is so obvious and visible in programmes of environmental education is the one enunciated by comte and reiterated by Pestcilogzi and Herbert Spencer, time and time again, that the education of the child must follow the same sequence as existed in the education of mankind, considered historically. This means that the genesis of knowledge in the individual should follow nearly the same course as the genesis of knowledge in the race. This principle can be followed both in letter and spirit in programs of environmental education. Children then taken out to a forest and made to comp there, can very well acquire, interpret and appreciate the sequence in which knowledge was developed by man, considered historically. To illustrate this point it may be said that when a child tooks at a brook and finds that some pebbles are pushing others and are being carried down the current, he sees for himself that flowing water has force which can be used for doing work or producing energy.

He can then interpret this knowledge when he goes to a hydro-electric power station and verify how water is used for producing electricity. Similarly, he can inter by feeling the wind how it is used as a source of energy for sailing of boats and barges. From all the aforesaid information, he may tru to raise a question for himself :' can he devise any method usingtidal waves for generating power and so on?"

7. The next important educational principle which is germane to the programmes of environmental education is the pleasurable excitement which these programmes create in the pupil. This principle can be appreciated by one who has seen the children's faces and the intense delight on them when they are picking up flowers and insects or hoarding pebbles and shells. If the adults do not thwart them, children may be led on the any

extent to conducting inquiry about the objects they collect. A teacher who takes children with him to the woods reports how reagerly the children help in searching out plants for him, how intently they watch him while he is examining these points and how they overwhelm him with questions. Later on they themselves collect plants, note to colours of the leaves and flowers, the forms of petals, the shapes and leaves, the number of the wings of insects, the legs, the antennae etc. Gradually they may learn to examine the stem, the pistil, the form of flowers whether radial or bilateral, the arrangement and the character of the leaves whether opposite or alternate stalked or sesile, s nooth or hairy, serrated, toothed or crenate, etc. Tne sum and substance of this principle is that whatever learning experiences the children are exposed to should be palatable to them and should also be within their com prehension and capacity. It is not so at present, for the child is taught many facts which do not come to it naturally and which do not serve any of its feltneeds. It has to memorise them to avoid failure in the examination. If forgets them soon after. Thus much of the time spent over memorising the facts mentioned above is a sheer waste of time. Note only that it creates in the child that distaste and aversion for learning which is reflected in its poor reading interests and habits. Since some of these children become teachers in later life, the lessons taught by them are poor in quality and monotonous in presentation. These shortcoming can only be removed if the beauty and glory of the environment is used as a teaching aid by the teachers and as a learning aid by the pupils.

8. The eighth principle of environmental education is that it makes child's education problem based, for understanding environment and the hazards of its pollution, the pollution of air and water, the destruction of wild life, the dereliction of land, etc are problems that all

of use should solve in order to save mankind from extinction. For illustration, the child should appreciate all the activities, such as Chipko movement which seek public co-operation for stopping the despoliation of forests and felling of trees and which are aimed at controlling air and water pollution, and should participate in them. In fact, he should develop such motivation for and commitment to solving environmental problems that he forms mohalla and city committees which can work voluntarily to stop people from dirtying their surroundings; committees which check people from excerting on roads and dumping garbage in their lanes, from throwing effuents into rivers; committees which put pressure on civic bodies for treating and disposing it properly; committees which check the playing of such vehicles on road which emit unpermissible quantities of smoke and carbon monoxide; committees which identify places where mosquitoes breed and clean them and spray pesticides; committees which impress upon thermal power stations to use filters for reducing the carbon mono-oxide and sulphur-di-odixe content of the smoke exhausted by them; committees which muster public support and efforts for solving their environmental problems and for improving their environment. Now working on these committees and doing all this is solving problems and using problem-solving approach for learning.

9. The last but not the least important principle of environmental education is its social relevance, its relevance to man's interaction with his physical and social environment, its relevance to changing human atitudes which cause man to hate man and beget hatred on one or the other ground, namely, colour, religion, region and caste, the hatred that precipaitates war, the hatred that we detest and yet it exists and explodes in the world in the form of Iran Iraq conflict. Arab-Israeli confrontation, Namibia's and SWAPO's problems,

skirmishes on the borders of various countries, etc. Such hatred, though it is at present confined to various regions of the world may sometime later trigger off a world war, a nuclear and chemical war which may cause the wiping away of mankind from earth. That would be a Doomsday. The environmental education curriculum by making students aware of this Doomsday and of the catastrophe and cataclysm that man's thoughtless behaviour causes is highly relevant socially. The environmental education curriculum is also socially relevant because it tells us how unchecked and unplanned development may pollute our air and water and erode our soil thereby threatening our subsistence and existence, it is thus obvious that environmental education is a socially relevant discipline and not something abstract or removed from the reality of our surrounding.

In conclusion, environmental education is a subject that is of a very practical nature and is also supoorted by sound pedagogical principoes. What its objectives are, is discussed in the next chapter.

The objectives of environmental education are part of the general objectives of education which have been defined variously by different thinkers, classificts. humanists and scientists Still most of them seem to agree with Herbert spencer that education should prepare for complete living, which does not mean living in the material sense merely but in the widest sense. That is, an educated person should know how to lead a healthful life, how to prevent one self from sickness, general debility, premature decrepitude, and short, how to treat the body; in what way to develop the intelect and to treat the mind; how to prepare for earning a living and to secure the necessaries and comforts of life; how to rear and discipline offsprings; how to maintain proper social and political relations; how to spend leisure time so as to gratify tastes and feelings; and in what way to use talent for the greatest advantage of the self as well as of others. For the accomplishment of most of these objectives, environmental education is an important medium. This is the reason why Zakir Husain wanted environment and productive work to be used as centre of learning. He said that there are three centres of correlation in Gandhiji's basic education; natural environment, social environment and craft work. These centres should be used to draw out the best in child and man.

Environmental education is indeed very important of child and adult for self-fulfilment and social development. It helps in the maintenance of life and health, in self-preservation and in the preservation of human race. It helps understand different food chains and the ecological balance in nature. It helps understand and appreciate how environment is used for making a living and for promoting material culture. It helps appreciated differed social institutions and regulate relationship in primary groups like family and secondary and tertiary groups like place of work, political institutions, etc. It helps in appreciating and enjoying nature and society. It stimulates concern for changing environment in a systematic manner for the maximum long-run as well as the immediate welfare of mankind. It directs attention towards the problem of population explosion, exhaustion of natural resources and the pollution of environment and sheds light on the methods of solving these problems. Many other benefit is of teaching environmental education can also be listed, as for example, the contentment and happiness one gets by looking at different gifts of nature or some other objects of environment for instance, some persons may become vary happy on seeing flowers, or on noticing a girl cutting and binding grain in a field and singing some melancholy strain. Some others may derive immense pleasure in reading about these things and in picturing mentally their images. This applies equally to social environment. A visit to a riot-torn area, or to one hit by a cyclone, or to one inhabited by unskilled workers increase, our sensibilities and susceptibilities towards unfortunate and less-privileged class of homo sapines

The objectives of environmental education can be subsumed in three domains discussed by Bloom in his book. "Taxonomy of Educational Objectives", cognitive, affective and psychomotor. The cognitive domain includes those objectives which deal with the recall or recognition of knowledge and development of intellectual skills and abilities, which means, it includes the following behaviours; remembering problem-solving; concept formation and to a limited extent, creative thinking. In other words, this area includes all conscious mental processes from ordinary recall or rerognition to higher ones, like solving a problem which involves abstract thinking or use of notations, or which relates to something so infinitesimal that it cannot be seen through even a very powerful microscope, like the orbitals of carbon atom. The affective do main includes the objectives that describe changes in interest, attitudes and values and the development of apreciations and adjustment. The area covers the entire continuum from ordinary attention to an object to deep own existence involvement such as that of Bhagat Singh, Chandrashekhar or Subhash Chandra Bose in the Indian freedom struggle, involvement the service of the suffering masses, involvement which is reflected in the indomitable spirit of those who conqure mountains and same rivers, who scud in space or stay as the bottom of the sea for days together. I. also includes contrary behaviour, that is, the aversion, antipathy or fear one shows for certain objects. The pshychomotor domain covers the manipulative or motor-skill area. This area includes neuromuscular coordinations found in handwriting, speech-making, performing physical exercises, dancing, doing yoga, winding a screw, using a saw and in doing a lot of the things which are required to be done under socially useful productive work and in vocational and technical courses.

The cognitive domain has been divided by Bloom into the following categories : 1. Knowledge, 2. Comprehension, 3. Application, 4. Analysis, 5. Synthesis, and 6. Evaluation. Knowledge has further been divided into knowledge of specifics (namely specific terminology and specific facts), knowledge of ways and means of dealing with specifics (namely, forms and conventions of the major types of work, *e.g.*, verse, plays, scientific papers; form and usage in speech and writing; structures in English language; style of speech, etc., trends and sequences, classifications and categories, criteria English language; style of speech, etc., trends and sequences, classifications and categories, criteria by which facts, principles and opinions and recreational activities are judged; methods and techniques of conducting inquiry); and knowledge of the intervals and abstractions in a field (*i.e.* principles and generalisation; theories and structures). Comprehension has been divided into translation (skill and ability to translate material into simple understandble language, and the ability to understand non-literary statements, metaphors, symbolism, irony, exaggeration) interpretation and extrapolation. Application means application of ideas, principles and theories, and the ability to predict the effect of change in a contributing variable on the phenomenon. Analysis means analysis of relationship and analysis of organisational principles. Synthesis means production of unique communication, preparation of a plan or dervivation of a set of abstract relations. Evaluation means judgements in terms of external criteria.

Bloom has further said that except the category 'knowledge', all other mentioned above constitute intellectual abilities and skills, and that both within categories and among them there exists a hierarchical order, from simple to complex, this is indeed very logical. So are the other principles used in preparing the taxonomy, namely that the objectives should be consistent with the distinctions teachers make among student-behaviours that they should be psychologically tenable and internally consistent and that they should include behavioural changes in a relatively neutral fashion. These are all laudable principles. Nevertheless, Blooms's classification of the objectives in the cognitive domain may not quite serve the purpose of delineating the objectives of environmental education which I have in my mind, for I want to highlight creative abilities (skills and abilities which relate to exploration, observation and discovery). By saying this, I do not mean that analytical ability or the ability to synthesis and evaluation are main components of thinking and reasoning ability, but is also includes some element of creativity and which also includes sieving and sifting of information that one possesses or gathers and organising it so as to give a more convincing alternative to the one under consideration, if there is any need to such an alternative. In other words, some element of creativity is also involved in every process of original thinking, reasoning and critical appraisal. This element of creativity is, in my opinion, weighty for a student of environmental education but has not been discussed or elaborated in the foregoing taxonomy. It is weighty in the sense that environmental education should help in searching new alternatives and new ideas for the prudent and sagacious use of the limited stock of environmental resources and sources of energy, and about reducing pollution, and also about making environment aesthetically, socially and culturally more enjoyable. This statement, however, does not mean that Bloom's taxonomy is redundant in respect of environmental education. It is useful in so far as the basic principles and the hierarchical order of the objectives are concerned. Nobody can deny that comprehension is not a higher mental process than the acquisition of knowledge or that application is not still higher, etc. similarly, nobody can deny the existence of a hierarchy from lower to a higher order within Bloom's each category. For examples, within the category 'Knowledge'. knowledge of theories and principles is certainly higher in cognition than knowledge of specific facts. This hierarchical order has got to be used in stating the objectives of environmental education also.

Now we can list the objectives of environmental education in all the three domains - cognitives, affective and psychomotor. The objectives in the cognitive domain are :

1. To help acquire knowledge of the immediate environment.

2. To help acquire knowledge of the environment beyond the immediate environment including distant environment.
3. To help understand the biotic and abiotic environment.
4. To help understand the effect of unchecked population growth or unplanned resource utilisation on the world of tomorrow.
5. To help understand the effect of unchecked population growth or unplanned resource utilization on the world of tomarrow.
6. To examine trends in the growth of population and interpret them for the socio-economic development of the country.
7. To evaluate the utilisation of physical and human resources and to suggest remedial measures.
8. To help diagnose the different causes of environmental pollution and to suggest remedial measures.
9. To help diagnose the cause of social tensions and to suggest methods for avoiding them.

Besides the foregoing objectives, the following skills and abili ies also fall in the cognitive domain :

(i) To help develop observational skills and notice details usually not seen by an untrained eye.

(ii) To help develop skills required for making discriminations in form, shape, sound, touch, habits and habitats.

(iii) To help develop ability to draw unbiased inferences and conclusions.

(iv) To help develop ability to make meaningful suggestions.

We can now list the affective objectives of environmental education.

1. To help acquire interest in the flore and fauna of the near and also distant environment.

2. To help evince interest in the people and problems of the community and society.
3. To show tolerance towards different castes, races, religions and cultures.
4. To appreciate the gifts of nature.
5. To love the neighbours and value mankind as a whole.
6. To value equality, liberty, fraternity, truth and justice.
7. To respect the national boundaries of all countries.
8. To value the cleanliness and purity of our environment.

Environmental education also fulfils some objectives of the psychomotor domain as are achieved through participation in environmental activities, excursions and camping programmes. Some of these objectives are listed below :

1. To participate in afforestation programmes.
2. To participate in programmes aimed at minimising air, water and noise pollution.
3. To participate in programmes aimed at preventing soil erosian.
4. To participate in programmes aimed at eliminating food contamination and adulteration.
5. To participate in cleaning neighbourhood.
6. To participate in urban and rural planning and execution programmes, such as, installation of gobar gas plants, solar heaters, etc.

The aforesaid objectives have been more or less delineated by the Unesco's Tbilsi Conference also, though in a nut-shell are quoted as below :

(a) The goals of environmental education are :

— "to foster a clear awareness of, and concern about, economic, social, political and ecological interdependence in urban and rural areas;

—to provide every person with opportunities to acquire the knowledge, values, attitudes, commitment and skills a needed to protect and improve the environment.

—to create new patterns of behaviour of individuals, groups and society as a whole towards the environment.

(b) The objectives of environmental education are categoised as follows :

Awareness : to help social groups and individuals acquire an awareness of and sensitivity to the total environment and its allied problems.

Knowledge : to help social groups and individuals gain a variety of experiences and acquire a basic understanding of the environment and its associated problems. Attitudes : to help social groups and individuals acquire a set of values and feelings of concern for the environment and the motivation for actively participating in environmental improvement and protection.

Skills : To help social groups and individuals acquire the skills for identifying and solving environmental problems.

Participation : to provide social groups and individuals with an opportunity to be actively involved at all levels in working towards the resolution of environmental problems.

In conclusion, comprehension of the objectives of environmental education is very essential for the successful formulation, implementation and evaluation of its programme. However, these objectives can only be achieved and understood properly, if we known what our environment is, what is contained in it.

The reason why we should choose a curriculum of environmental education is that we are both and brought up in an environment; we eke out our living in an environment; and whom the sister death pinions us and we breathe out our last, we do that too in an environment. Thus, from conception to cremation, we live in some or the other environment, and use it for meeting all our needs, comforts and luxuries. We should, therefore, understand our environment and know how to use it for our individual well being and for the welfare of the society and the future generations. Thus environmental education is more functional than the disciplines wer have been studying so far.

Indeed disciplines such as Physics, Chemistry, Zoology, Botany, Geography, History, Economics, etc. should all be the means to promote the immediate and future welfare of mankind and they should not be an end in themselves. But it is not actually so, as we teach these subjects in most schools to a very great extent for their own sake. Children learn them, I memorise the facts about them, without being able to comprehend as to how they would use them later in their life. Thus they are lost in subterfuges and means and miss the end. For instance, they learn the Mathematical definitions, memorise long structural and molecular formulae, learn how complicated compounds are prepared in the laboratory, etc. but they do not know the names of the trees of their neighbourhood, the trees they should grow in the school to beautify their campus, the trees they should grow for fuel and timber, the trees they should plant along boundary walls, the structure and potential of employment in their neighbourhood, the cottage and small scale and large scale industries that can be started in their locality, the institutions whereform and the purposes for which they can taken loan, etc.

Thus discipline approach in science and social sciences is not functional at school and even college level. It merely prepares students for receiving higher education. It is, therefore, meaningless for majority of the students who drop our before our after completing the primary classes, or discontinue their education in the middle or high school. It does not even sharpen their thinking and reasoning ability, because most of the books available in the market lay emphasis on rote memory, a play 'Refund' by Fritz Karinthy illustrates this point very well. A man named Wasserkopf addressing the principal of the school where he had studied, said, "As a former pupil of this school I want you to refund the tuition fee which was paid you for my education eighteen years ego" The principal asked him, 'Why do you want it back ?" He replied, "Because I didn't get my money's worth, that's why. This certificate says that I got an education. Well, I didn't. I didn't learn anything You took muy money, and you taught me nothing. Now I'm no good for anything and I can't do the things I should have learned in school. I was given instruction here in exchange for money so that I might be able to do something, but I can't do anything." Though these lines have been written jokingly and the play is not meant to prove the uselessness of the instruction that is given in schools, yet it does it. It proves that the teaching done in schools is not related to life and environment. This explains the need and asers the "Why" of the curriculum of environmental education.

The next question is about the objectives of environmental education, for it is upon the objectives that the contents of the curriculum would depend, the objectives have been given in Chapter 3 (supra). These objectives can obviously be achieved by following either of the two alternatives :

1. By treating environmental education as an approach and teaching it as a part of Science, Social Studies, Languages, and other subjects; or
2. By treating it as a separate subject.

Alternative 1 is no doubt very lofty and loudable but it does not provide environmental education that structure which is requird for sequential development of knowledge and for meeting the needs of a spiral curriculum. As against this, Alternative 2 provides for a definite shape and structure to the subject of environmental education but it deprives it of its holistic character which is germane to Alternative 1. Alternative 2 has been assigned the name, 'Infused Model' in the Discussion Guide for Unewsco Training Workshops on Environmental Education published in 1980. This model is also called interdisciplinary model, for it is intended to draw out units from different disciplines and to put them into one namely environmental education. The other model is based on Alternative 1 and is called infusion model in which environmental education will be taugh as a part of various subjects.

Whatever model we may adopt, the objectives and contents should be selected keeping in view the developmental needs of the children and the welfare of the society. The developmental needs of the children have been discussed at length by many a psychologist. Piaget deals with those subsumed under intellectual development, the main forte of educators. He identifies four successive stages which he believes are child's principal means of knowing his environment. One is sensori-motor stage, from birth to two years of age; second is pre-operational stage, from the beginning of the third year to the end of the eighth year; third is concrete operational stage, from the beginning of the ninth year to the end of the twelfth or thirteenth year; fourth is the stage of formal operations, from the thirteenth or fourteenth year onwards. Of these four stages, we would be concerned with the second and third only, as we would concentrate on children who range in the age group, five to thirteen. The children in the second stage are always in sort of a wonderland. Their cognition is native, impression-bound and poorly organised. They look at things with a feeling of wonder, examine them for some time and then discard them. They may explore things, collect a few of them, keep them for some time and then throw them awaqy.

It is this desire for exploration and collection that the teachers of environmental education can depend upon and use as advantage position for encouraging children collect things of their choice. Some may collect leaves, some others pebbles, still some others wild figs, etc. The teacher may then discuss these collections in the class and ask children to collect a few more pieces of the specimen collected by them earlier. This may help children in sharpening their recognition skills which are the basis of reasoning and the foundation of human thought. During their next stage of development when children are usually in class IV to VII they will rely upon concrete objects to facilitate the logical organisation of their cognitive processes and are unable to deal successfully with verbal expressions and abstractions. This does not in anyway contradict Bruner's postulate that it is possible to introduce children at any age to ideas and styles that in later life make an educated man. To illustrate this point, it may be said that the children may focus their attention on their own house till them are eight or nine years old that is, till they are in class I to III, but they should revisit the topic through the concept of 'shelter' when they are ten or eleven years old, that is, in class IV or V. This sort of revisiting the topic is reflected in the concentricity principle and the 'spiral curriculum' of Bruner.

Keeping the foregoing considerations and the objectives of environmental education, the following activities can be undertaken by students of environmental education :

1. Observing plants with a view to recognise them.
2. Observing flowers with a view to recognise them.
3. Observing birds and animals with a view to recognise them.
4. Discriminating between colours and sizes of different plants, flowers and animals and learning their names.
5. Hearing and imitating different sounds.
6. Counting the articles collected, calculating the number of Khurpis needed to dig ground for planting saplings.
7. Observing changes in growing plants and animals by making comparisons and measurements.
8. Building hut and houses with blocks.
9. Measuring the area of flower and vegetable beds and investigating spatial relationships.
10. Discussing about the observations made.
11. Writing descriptions of and reactions to the things observed.
12. Drawing pictures of trees, hills, houses, etc.

2. To help acquire knowledge of the environment beyond the immediate environment including distant environment.
3. To help understand the biotic and abiotic environment.
4. To help understand the effect of unchecked population growth or unplanned resource utilisation on the world of tomorrow.
5. To help understand the effect of unchecked population growth or unplanned resource utilization on the world of tomarrow.
6. To examine trends in the growth of population and interpret them for the socio-economic development of the country.
7. To evaluate the utilisation of physical and human resources and to suggest remedial measures.
8. To help diagnose the different causes of environmental pollution and to suggest remedial measures.
9. To help diagnose the cause of social tensions and to suggest methods for avoiding them.

Besides the foregoing objectives, the following skills and abili ies also fall in the cognitive domain :

(i) To help develop observational skills and notice details usually not seen by an untrained eye.

(ii) To help develop skills required for making discriminations in form, shape, sound, touch, habits and habitats.

(iii) To help develop ability to draw unbiased inferences and conclusions.

(iv) To help develop ability to make meaningful suggestions.

We can now list the affective objectives of environmental education.

1. To help acquire interest in the flore and fauna of the near and also distant environment.

ploration, observation and discovery). By saying this, I do not mean that analytical ability or the ability to synthesis and evaluation are main components of thinking and reasoning ability, but is also includes some element of creativity and which also includes sieving and sifting of information that one possesses or gathers and organising it so as to give a more convincing alternative to the one under consideration, if there is any need to such an alternative. In other words, some element of creativity is also involved in every process of original thinking, reasoning and critical appraisal. This element of creativity is, in my opinion, weighty for a student of environmental education but has not been discussed or elaborated in the foregoing taxonomy. It is weighty in the sense that environmental education should help in searching new alternatives and new ideas for the prudent and sagacious use of the limited stock of environmental resources and sources of energy, and about reducing pollution, and also about making environment aesthetically, socially and culturally more enjoyable. This statement, however, does not mean that Bloom's taxonomy is redundant in respect of environmental education. It is useful in so far as the basic principles and the hierarchical order of the objectives are concerned. Nobody can deny that comprehension is not a higher mental process than the acquisition of knowledge or that application is not still higher, etc. similarly, nobody can deny the existence of a hierarchy from lower to a higher order within Bloom's each category. For examples, within the category 'Knowledge'. knowledge of theories and principles is certainly higher in cognition than knowledge of specific facts. This hierarchical order has got to be used in stating the objectives of environmental education also.

Now we can list the objectives of environmental education in all the three domains - cognitives, affective and psychomotor. The objectives in the cognitive domain are :

1. To help acquire knowledge of the immediate environment.

Twentieth-Century Analytic Philosophy